The Voodoo Vortex

By

Rus Morgan

This book is a work of fiction. Places, events, and situations in this story are purely fictional. Any resemblance to actual persons, living or dead, is coincidental.

ISBN: 1-4033-4258-X (e-book)
ISBN: 1-4033-4259-8 (Paperback)
ISBN: 1-4033-4260-1 (Dust Jacket)

This book is printed on acid free paper.

1stBooks — rev. 04/03/03

Acknowledgement

My profound thanks to the professionals in the ER and the Cardiac Ward, in particular Linda and Michela and all the rest of the staff at the Veteran's Hospital in Memphis without whose care, concern and life-saving techniques this novel would never have been published. My gratitude to all those close to me—whom I appreciate more each day. To my wife Vonn for understanding that life is enriched and books are written because loved ones allow you the time to create. My very special thanks to my son Brad for applying his patient, insightful perspective as he picked the specks out of the pepper—and for rendering informed, pertinent comments where applicable.

Dedication

Dedicated to my son

Bradley William Morgan.

'Tis sad there's only one of him.

In memoriam

G.E.M.

1931 - 2001

OF NOTE...

This is the first in a series of suspense novels featuring these three intriguing characters. Jason Amador, sixty year old, reclusive, multi-millionaire genius with an IQ in excess of 180, speaker of six languages. Possessor of a photographic memory with total recall.

His personal assistant the beautiful, dusky, Jamaican ex-hooker. She of the startling blue eyes, who saved the money from her tricks and rewarded herself with a Master's Degree in Computers. She programs 'Cray' Amador's unique triple mode computer—so fast it needs an icy mountain trout stream constantly running through it to siphon away the heat. Amador's muscle is supplied by tall, handsome, ex-con Peter Chaney who enjoys killing but now only kills in defense of Jason Amador. Chaney taught himself the law during the ten years he was locked up for murder. He has the equivalent of a 'black belt' in four different martial arts disciplines and craves Christa Bonamay who is resistant to his charms.

CHAPTER 1

The desperate figure came flying over the fence and turned into a human cannon ball as he hit the ground. He did two complete rolls then unfolded his short, serpentine bulk and melted into the sparse forest surrounding the fence. It happened so quickly and quietly that if you had been looking any place but there you would have missed it.

He sprinted a hundred yards away from the fence, made an abrupt stop and quickly unrolled a bundle tied to his waist. It was a camouflaged poncho with an inner layer of foil he had laboriously hand stitched to the lining. He was gambling it would conceal the heat of his body until he got out of range of the infrared sensors. Under the waning light of a full moon it was no trouble to find the foot of the deer trail and he began his tortuous climb to freedom.

* * * * *

"Commander."

The low voice was urgent enough to raise him from sleep. "Yes?"

1

"A million pardons, Honorable Commander but we just finished 3 a.m. bed check and G-101 is missing."

Senior Colonel Pak Pong-Yul tweaked the light switch and confirmed the time. "Is he in or out of the building?"

"Definitely out, sir."

"That is confirmed?"

"Yes sir, the alarm system was compromised and the door was forced."

"How about the grounds?"

"I have three squads checking the compound now, sir."

"Anything on infrared?"

"Negative"

"That's curious, how about the rest of Dr. Ho-Sik's subjects?"

"Everyone else accounted for, Sir."

"I shall be there shortly."

"Yes Sir. Does The Commander wish me to sound a general alarm?"

"No."

Senior Colonel Pak Pong-Yul of the Strategic Forces Command, North Korean Peoples Army, (NKPA) Commander of the special covert detachment at Voodoo Mountain, shrugged into his pants and checked himself in the mirror. At forty-two he had been hand-picked for this assignment deep within the enemy's own bowels. The bastard son of an unknown American GI and a Korean housemaid, the very size he had inherited from his father allowed him to dominate his kind in the streets but was the same size that killed his mother in unassisted childbirth. He had managed to steal enough to get himself educated and off the streets of Pyong-Yang where his peers still fought and scratched for each meal. In school and later in the military he had been outcast, friendless, quick to avenge the slurs and slander evoked by the hated American blood. His size, quick temper and readiness to fray eventually guaranteed his privacy and commanded grudging respect from those around him. He had received highest marks in his studies at the university and also later in the elite Officer's Cadre where he had excelled in most subjects including the study of the English language. The army was his home, his mother and his father.

It had been good to him and as he shrugged into his coat his chest full of medals bespoke his total devotion to it.

He reached for the light switch and paused, his hand poised above a picture of a woman and two small boys. She was smiling up at him and the camera had caught the two urchins in mid giggle. He picked up the photo and tipped it toward the light. He had first seen her on one of his field trips into the mountains. A rare mountain beauty, graceful as a young gazelle. He had made overtures to her father and soon a price was struck and he bore her proudly back to the city. She had given him two fine sons and he missed them both. They were excelling in their studies and would honor him by following in his footsteps. He smiled briefly back at the photo. Only he knew that he missed her more than he missed the boys. He slipped the picture from the frame and slid it out of sight in the left side of his tunic.

He strode briskly through the door into the glistening, darkened corridor. In two minutes he pushed through the door into the command center.

Watch Commander Lieutenant Kye Pong-Uk saluted him and stood stiffly at attention waiting for his Commander's word.

The Commander gazed intently at each security monitor in turn. He saw the three squads as they probed the compound. "Stand down, Lieutenant. Do you have any results?"

As he spoke the question the Squad Leader of the perimeter fence detail stopped and addressed them by radio. Lieutenant Pong-Uk fiddled with a knob and brought the squad into zoom focus on a monitor. The speaker's face was clearly visible in the floodlit compound. He was pointing to the fence. "Watch Commander, he went over the fence here."

The Colonel spoke into the microphone. "Sergeant, this is your Commander. Talk to me."

The figure on screen came to immediate, rigid attention. "Honorable Colonel Sir, there is an extra long tripod leg lying here in the grass…"

"Explain yourself."

The soldier bent down then lifted a shiny pole into the air so the camera could see it. "Sir, it looks like he put three tripod legs together and vaulted the fence, sir."

"Vaulted the fence?" That goddamned fence is eight feet plus razor wire. "What makes you think he vaulted it?"

"No damage to the fence Honorable Commander sir. The fence is hot but he's definitely not inside the compound sir."

"You are two hundred yards from the North Gate but you can see it from there, correct?"

The Sergeant shaded his eyes against the floodlights. "Correct sir."

"Proceed to the North Gate. Go outside and sweep the caldera quickly. If you find him bring him to me…alive if possible."

"Yes sir."

Pong-Yul turned to the Watch Commander. "Pass them through the gate and give them thirty minutes to complete their sweep. When did Blue Squad return from outside recon and how long was that fence turned off when they came in?"

The Watch Commander flipped his log book back a page. He pointed to it with obvious relief, "Forty-nine seconds sir. They came in exactly one hour and forty minutes ago."

"So he's been out there an hour and forty minutes already. Could you vault that fence in forty-nine seconds, Lieutenant?"

"No sir."

"Neither could I, but we're dealing with a desperate, very strong martial artist in G-101. He could, and did. The tripod must have snapped back off the fence just before the electricity flowed again. We have to get him back, dead or alive."

"Yes sir."

"What are you picking up on infrared sweep?"

"Nothing large enough to be a man Sir. I've got some rodents," and he pointed at the specks on the screen, "and three probable deer up on the opposite ridge," he zoomed in on three large pips on another screen." They are too far away and too high for G-101. He wouldn't have had time to make it that far up yet."

"He's found some way to fool our sensors. Break out a two-man Chopper." The Commander said. He paused for a moment, surveying the monitors. "The flight plan is this. There is only one main path out on that side. The pilot is to go immediately to it and inspect it to the rim. If he doesn't find G-101 than he is to come back to the valley floor and sweep back and forth for a hundred meters on each side of the path and do that in fifteen meter layers until he finds the traitor. When he finds him I want him forced off the wall. But don't terminate him unless there is no other way."

He stepped to the map on the wall and studied it. "If we don't find him before 0500 I want outside recon." He pointed to the areas. "Put out four squads of four. One north, one east, one south, one west. Permanent floating duty, no relief, rations for a week. That should give each squad time to dissect their areas. I want him back dead or alive."

He strode toward the door then stopped and turned. "Lieutenant, send our best squad leaders. They are to remain covert at all costs. However if they are discovered they are to kill those who have discovered them and make it look like an accident. Let me know when they're ready and keep me advised of any news. Now call Dr. Ho-Sik and tell him I'm on my way. I wish to go over his security measures."

CHAPTER 2

It looked like an agile child who ran effortlessly up the dim path towards the crest. On closer inspection however the only similarity to a child was the short stature. As his stubby legs propelled him up wards sweat poured from his head into his shirt and down both back and front. The heat was suffocating under the cloak but he ignored it and continued to put more vertical distance between himself and the lighted compound below.

Abruptly the grounds below lit up like day and three squads of four soldiers each broke into the open and began probing the compound. He watched them for a moment as they scurried about like ants looking for interlopers. One squad congealed at the point where he had vaulted the fence and he saw a tiny figure hold up something long and bright. His pole had been discovered.

His frenzied climbing had now brought him almost to the half way mark but he still had nearly three hundred meters to go when a low whirring sound below him made him pause and study the compound. It lay like a contractor's model some nine hundred feet below him.

The perimeter fence ran approximately three hundred yards on a side and two hundred yards across the one end. Floodlights along its length and strategically placed within the compound now made the whole available to his scrutiny. Three small buildings were visible within the compound. Each was painted in green camouflage and nestled within scraggly groups of well kept pines. He knew that certain of those pines were dummies and harbored anything from air vents to cameras and sensors.

The three squads had converged on the north gate and were now flowing through and beginning the inevitable check of the valley floor.

The other end of the fence butted abruptly into the rocky face of the Caldera which rose perpendicularly some two hundred feet through the mists above the valley floor and disappeared into rocky snaggles covered with scraggly brush. This rocky face was where the sound came from. As he watched part of the rock at the valley floor shifted and began to slide sideways.

"A Chopper," he muttered as he pushed his weight up the winding path. A two man chopper came through the opening in the rock face like a giant dragonfly. It rose quickly across the fence where he had made his vault then swung to the left and hovered just out of blade range of the wall. The pilot picked up the beginning of the path three hundred meters below him. A powerful beam lanced from the chopper painting the path a brilliant white. Like a yo-yo on its upward climb the chopper began to rise rapidly searching the path with insidious curiosity.

G-101 broke from the trail, darted along a level shelf for twenty meters then started a nearly vertical climb up the wall. His fingers closed on handhold after handhold and the muscles in the short powerful arms rolled and bunched till his shoulders bulged like a small hunchback. He picked his way up the face of the caldera like a fly on the wall. So perfect was his camouflage that he was just an irregular interruption in the moonlight reflecting from the mica. As the chopper rose level with him he flattened himself against the rock and held his breath. The bird continued to rise with the path and went to the top where it hovered for a moment then abruptly dropped back into the caldera and down to the valley floor where it began to glide slowly parallel to the ground until it passed below him and continued on for a hundred meters—then turned back and repeated the pattern twenty meters higher.

He knew the pilot would maintain that search pattern straight to the top of the caldera. He estimated each swing would take fifteen minutes. He had less than an hour and then the vicious glare of the searchlight would pin him to the wall as a moth against a windowpane. He began climbing with renewed purpose.

CHAPTER 3

Pong-Yul strode forcefully into Doctor Ho-Sik's small office. He had great respect for the Doctor's achievements and would rather have broached this matter at another time but G-101's escape left him no choice. He made just enough of a bow to preserve protocol but not enough to ease the tension of the moment.

He spoke quickly. "My apologies for dragging you out of bed at this hour Doctor, but you are aware of the situation I assume."

Doctor Ho-Sik shrugged and the shoulders of his perennial white smock raised slightly. "It is of no matter, I have been up since *we* discovered the escape."

"Yes, tell me exactly what happened."

Ho-Sik pulled a pack of cigarettes from his pocket, offered one first to Pong-Yul, who declined, than took one and lit it.

He held the cigarette in the classic oriental palms-up position between forefinger and thumb, took a deep drag and squinted through the smoke. "My staff is on two twelve-hour shifts beginning at 8 am. I work from 8 am as late as necessary and am on call twenty-four hours. Three quarters of my staff are on the day shift where we have the most interaction with the subjects. During the day shift we are all involved with the subjects so no checks are necessary."

He paused to let this sink in. The Commander waited but not too patiently. "The night shift from 8 pm to 8 am is active up through midnight so again no bed checks are necessary." He waved his hand in the air and the smoke danced in small vortexes around the two men.

Ho-Sik continued. "All of these subjects are volunteers and have dedicated their lives to the 'Central People's Committee' and to this project. Even so, at 3 am we make bed-check. At that time the orderly discovered the alarm had been disabled and the lock had been forced. We knew someone was missing but it was a few minutes before we discovered it was G-101."

Pong-Yul cut in. "Has he given you any reason lately to suspect he might be going to defect?"

"Not directly, no. As you know he was my first human subject and only a partial success. What you do not know is that he is slowly dying."

This news stung Pong-Yul. As Commander he should have been notified. He hated having untrained civilians under his command and especially loathed this insolent little doctor. "How long does he have?" he grated.

Ho-Sik ignored the other's tone. He did not like this tall, bastard Colonel regardless of his accomplishments. "Six months—perhaps less."

"He knows this?"

"Yes."

"How long has he known?"

"We determined that point last week and after the initial shock he seemed to accept it the way any good party member would."

"He's very strong and extremely agile."

"He will be for some time yet."

Pong-Yul returned to his original mission. "How did he slip out of here this morning?"

"In deference to his condition I prescribed sleeping pills instead of injections to help him sleep at night. I authorized 30 milligrams of Restoril last night." He picked up a log book, turning several pages, then ran his finger down a column. "He was given two fifteen milligram capsules at 20:00 hours. That should have put him to sleep for the rest of the night. He must not have taken them, however he was observed in the bed at midnight"

Pong-Yul listened to the doctor but was reasoning quickly in his mind...G-101 has been out for at least two hours now...with this bright moon he's got to be near the top...

He spoke, "A costly error in judgment, Doctor."

Ho-Sik's small frame went rigid. He leaned forward and skewered Pong-Yul through his glasses, angrily spitting out his words. "Neither you, nor anyone else on this earth has the right nor the qualifications to judge me in error Colonel. The work I am doing here is breaking radically new ground in the medico/scientific world and no one, I repeat, no one else is capable of it..."

Pong-Yul bowed his head in slight deference but his tone was sharp. "I know Doctor Ho-Sik and I sincerely appreciate your genius but we've got a loose cannon running around outside now and I want to make sure he's not joined by anyone else. This whole mission could be in jeopardy..."

Ho-Sik snorted. "The rest of these volunteers are dedicated party members whose fealty is beyond question—no fence is necessary to keep them…"

"I would have thought G-101 was in that same category."

"He was but he evidently lost purpose shortly after he learned his condition was terminal."

The Colonel shrugged. "We're all terminal—and expendable."

The Doctor stubbed his cigarette viciously into the ashtray. "Yes, but we won't die the way he will. Most of the time he feels O.K. but then the altered DNA I infused comes active and for a short time his body literally flows and he's in great pain. Without sedation and pain medication he can be a very dangerous madman during those episodes. Eventually his viscera will be like bad rice and he'll turn into an inert blob with no heart, lungs or other discernible organs."

Pong-Yul whistled gently. "All the more reason I've got to catch him…."

A crackling speaker on the small radio attached to his shoulder cut him short. "Honorable Commander"

Pong-Yul toggled a switch. "Yes."

"Sir, all squads reporting from the valley. No sighting of any kind."

"Still nothing on infrared?"

"No sir,"

Pong-Yul glanced at his watch. It read 03:57. "What's the status on the four recon squads?"

"Assembled and awaiting your orders, Sir."

"I'll speak to them in five minutes, I want them clear of the North Gate before first light."

"Yes Sir."

CHAPTER 4

G-101 wiped a stubby forearm across his forehead. The temperature was unbearable under the camouflaged poncho which draped him from head to toe. Sweat ran in rivulets from his forehead to his chin and his clothes were soaked but he dare not come from under the protective covering. He knew it was functioning as a block for his body heat because the chopper had not come directly to him.

Now it was directly below him about sixty meters and climbing steadily and he was still fifteen meters from the top and beginning to tire. His bloodied hands were cramping, threatening to relinquish the tenacious grip that had brought him so near the top and freedom. His head was spinning from the heat and his body was violently complaining with each agonizing handhold. I must cool down, I must cool down, he thought, but I must not let the chopper spot me.

He decided to chance an all out climb. He slid from under the poncho and renewed his scrabble up the steep slope in a desperate bid to make it over the top before being spotted from the ground or the chopper.

* * * * *

"Commander!" The tone was respectful but insistent.

Pong-Yul had just given his troops their final orders and was returning via the hall to the Command Center. He touched the button on his shoulder radio. "Yes?"

"I've spotted G-101 sir."

"I'm on my way."

He pushed through the door of the Command Center and began searching the monitors. "Where?"

Lieutenant Pong-Uk came to attention. "He just burst over the top Sir. He wasn't there then suddenly he broke just above where the chopper was searching and scrambled over the top and out of sight."

Pong-Yul punched buttons on the consul then stood there watching an instant replay of the last thirty seconds on the screen. There was nothing but blackness above the bright red spot of the Chopper's exhaust then suddenly a point of red appeared above it and moved rapidly up screen and abruptly disappeared. The screen flicked back to real time and the Pilot's voice came in.

"Lieutenant Pong-Uk?"

Pong-Uk toggled a switch. "Yes"

"My orders from The Commander were specific sir, I was to stay inside the caldera. What are my orders.?"

Pong-Uk arched his eyebrows at Pong-Yul. The Colonel took the mike. "Pilot, this is Colonel Pong-Yul. Proceed outside and try to find him. Update me every five minutes."

The chopper swept off screen and Pong-Yul turned to the Watch Commander. "Contact the squads and give them G-101's last position. Tell them to force march up to there. He's got to be exhausted by his climb. If the chopper can find him and pin him down we'll get him yet."

* * * * * *

G-101 came out of the caldera on a flat spot sparsely covered with grass and rocks. He bent over from the waist and sprinted to the left of his position. With his legs pumping and his heart pounding he had covered the better part of two hundred meters when the chopper popped over the edge of the caldera and began to sweep with its searchlight. He dropped to the ground and watched it dip over the outside edge and slowly began to probe the hillside immediately below.

He heard shouts from below and peeked over. The north gate was directly beneath him and four squads of ant-like soldiers were pouring through and double-timing toward the path below him. He grinned through the ache in his chest and began to lope east along the top of the caldera. One hour would put him out of reach of most of the pursuit.

CHAPTER 5

"Benjie—you guys ready? If you're gonna be on Number 1 by first light you'd best shake a leg."

Charles Burlingame, in pajamas and dressing gown, stood at the foot of the winding staircase in the Burlingame family mansion. A stocky millionaire widower: one time football player at the University of Georgia; outstanding graduate of the Chicago School of Law; owner of a bank, a hardware store, three grocery stores, two shopping malls, fifteen convenience stores and several thousand acres of prime land; father of Benjie, short for Benjamin and his younger sister Kathleen; was impatient. He wanted to go back to bed and get another two hours sleep.

"Yeah Dad, we'll be right down."

Burlingame checked his watch. "It's 4:45 now. The Jeep's out front, leave the keys at the store with Kathleen and pick 'em up when you come back out. See you when you get back. Be careful."

Benjie winked at Harlowe Chapman jr., his dorm mate at college and best friend. "We'd better move Chap, Pop's about to have a

23

nervous breakdown." The two young men shouldered their light packs and glided down stairs, their hiking boots softly thumping each step. Blond haired "Benjie" and dark haired "Chap" both moved with the fluid grace which is the preserve of young athletes whose muscles are trained for maximum performance but who luxuriate when they get the rare opportunity to loaf. They disappeared out the front door in a haze of chatter and, from the bedroom, Burlingame listened to the Jeep's engine drone off into the pre-dawn night.

* * * * *

The northeast Georgia dawn was the kind you dream about. The moon stayed up long enough to watch the sun burn the haze off the high meadows while wary white tailed deer caught their last few mouthfuls of the succulent grass. In the bottoms along Highway 115 the sun had not yet cleared the mist and the jeep pushed its way through the swirling moisture as the boys made their way to the store.

Benjie parked the Jeep in a corner of the lot and waved at a comely, short blond outlined in the front door of the store. Then the two boys crossed the two lane road and faded into the diminishing mist. They would walk and jog approximately eleven miles up hill through the woods past the Crevasse to the end of Trail #1 then travel due east cross country north over to the edge of the caldera. Then around the south side of the caldera to the beginning of Number 2 and South on #2 for thirteen miles back down to Highway 115 and Southwest to the store and the jeep. It was an ambitious hike but both youngsters relished the exercise as a way to keep in shape.

About three quarters of a mile above the store they threw themselves down on the banks of Flint Creek and cooled their faces, had a noisy water fight and continued the walk up the hill. The cool, overhanging path above the creek was peppered with deer sign and they hoped to see the regal whitetails so they moved in silence. The trail straightened out to lead up to the old logging road and they broke out into the first of a series of high meadows with thigh-hi grass. They surprised an eight pointer and three Does at the upper end. The buck, acting from instinct and self preservation snorted, flashed his white tail and bolted into the safety of the woods. The Does stood firm, merely raising their graceful heads and pointing their ears as the boys walked on.

CHAPTER 6

A faint blush of eastern gray began to dilute the blackness in the sky. The stars retreated, grew dim then disappeared as the first rays of the sun tinged the gray with red. The start of another perfectly gorgeous day.

Most of this was lost on the grotesque little figure hunched in an exhausted lump against a large rock on the southwestern rim of the caldera. He opened his eyes in the early morning light, listened for a few moments without moving then glanced furtively around him. Sensing nothing threatening he stood, stretched and moved rapidly down the outside slope of the caldera. There would be peace and quiet down below at least for a little while and somewhere some food and water. He smiled at the thought as he entered the cool safety of the high woods.

* * * * *

Benjie and Chap had little real distance to travel for such accomplished hikers and all day to do it. Had they continued straight up Trail 1 as fast as they could jog they'd have reached Voodoo Mountain in late morning. Instead, when they came out on the old overgrown logging road they decided to turn to the left and go north for a while. It would make them a little after dark before they got back to the jeep but who was counting?

In about an hour they came upon a small meadow and while crossing it to the east kicked up a covey of six Bobwhites. The birds were gone in a bobbing, scolding flash of wings. The boys re-entered the woods, picked a cool spot free of ants, ate their lunch and enjoyed a twenty minute nap. After burying their debris they set a course Northeast by North with the aid of Benjie's pocket compass and in early afternoon rejoined Trail #1 and jogged northward.

* * * * *

G-101 felt a stab of pain in his gut and sat down abruptly. It was the first one he'd had in almost a week. He grunted and leaned far over his knees trying to ease the agony. A searing, wrenching tremor coursed from his belt line to his backbone and made him suck his breath through clenched teeth. He began sweating profusely and slumped over on his side with his stubby arms clasped around his

27

middle. The world spun around and he was enveloped in a hazy red fog that soon turned to merciful blackness.

* * * * *

The hikers had crossed the cutoff to the Crevasse, turned east and were racing each other down the brow of the mountain through the trees towards the juncture with Trail #2. As they flew down the slope the latticed rays of the mid-afternoon sun nearly blinded them. Benjie, in the lead but closely followed by Chap was bolting downhill. It was exhilarating to combine gravity with headlong rambunctious flight to leap over bushes, fallen logs and rocks. He came to a small copse and cleared it with a leap. In mid-air he noticed something childlike nestled against the down-slope side. He gave a shout as he landed and took several stumbling steps before he could slide to a halt. Chap was quick upon him and the two went sprawling in the pine needles.

"What's with you?" Chap demanded wiping needles off his face.

"There's someone lying back up there."

"I didn't see anyone," scoffed Chap.

"Look back up there," Benjie pointed. "You can see his pant legs from here."

"Well, let's go take a look, I hope he's all right."

CHAPTER 7

"Let me speak to the Sheriff?"

"Whosis?"

"Charles Burlingame."

"Yessir, Mr. Burlingame, right away."

"Sheriff Gibbs here."

"Wayne, Charles Burlingame—sorry to bother you at this time of night Wayne, but I'm glad I caught you."

The sheriff put his pipe down carefully in the pipe rack at the back of his cluttered desk so he could give full attention to his caller. It wasn't often the richest man in the county called you on the telephone at 10:30 at night.

"Jest tyin up loose ends, Mr. Charles. Whut kin Ah do for you."

Burlingame drew a deep breath and tried to sound nonchalant. It didn't quite come off. He sighed. "Wayne, I'm probably just an anxious father but I want to know if you've heard anything 'bout hikers problems today."

"No sir, we haven't. Been most quiet lately. What's the mattah?"

"Well, Benjie—you know my boy…"

"Yessir, I know him."

"Benjie and a college buddy of his took off on a hiking trip up 'round the south side of Voodoo Mountain this morning. Should have been back around dark."

Sheriff Gibbs looked up at the four by five foot map on the wall across from his desk. "Which way'd they go up?"

Burlingame's voice was hollow, now filled with worry. "They left the jeep at the convenience store on 115 just before sunup. They were going up on #1 to the head and cross over and come back on #2. That's a long way for you and me but the boys figured they could do it sunup to sundown."

"Well—Benjie's an experienced hiker, how about the other?"

"Strong youngster, on the football team."

"The only thing I can do right now, Mr. Charles, is send a jeep up the old logging road with a good searchlight on it. Ah'll dew that in the next twenty minutes. Tew a my deputies like to run for fun so at first light Ah'll put one a them up Trail One and the other up Trail Tew. We aught to have them boys located by 7 a.m. if not afore."

"Thanks Wayne, you tell those deputies I'll personally pay them double overtime."

"That won't be necessary Mr. Charles and I wouldn't be worried about them boys. Yew know how young'uns is. They're sittin aroun a campfire someres up there jist havin a good ole time."

"I hope you're right, Sheriff. Let me know as soon as you find out anything."

"Ah'll call you first news I get, Mr. Charles."

"Thanks. I'll be here at home."

CHAPTER 8

Caucasians have such large, thick bodies, he thought. If these two were my people they would be much easier to move. My ancestors are smiling on me by providing me with moonlight so that I may see to finish my labors.

* * * * *

The jeep second geared its way up to the end of the old logging road, loudspeaker blaring and a roving searchlight turning the surrounding forest into light and shadows but there was no answer from humans. The deputies spotted many eyes in the forest but all faded silently back into the woods.

By seven a.m. the next morning both runners had reached the ends of trails 1 and 2 and had joined each other in between but there was no sign of the hikers.

By noon Sheriff Gibbs had tapped all his two-legged resources and had them fanned out on both Trails 1 and 2 and the old logging road plus the more unlikely places like Trails 3 and 4 and the second old logging road that ran up behind Lake Pawnoo. There were twelve summer homes on the south and east side of the lake, seven of which had occupants who were questioned but had seen nothing.

The Sheriff's department unofficially boasted one track dog—Old Petey. Although he was the personal pet of Deputy Orvil Vaughn, Old Petey had an established reputation as a tracker having led his master to more than one lost hiker around Voodoo Mountain. Orvil was supplied with one of Benjie's other boots and after the long-nosed hound got a good smell of the foot gear he took his master straight across the road and up Trail #1. Petey would like to have sprinted along the scent but Orvil held him to a more comfortable gait and it was nearly sundown when Petey led him back onto Trail 1, went north for a while and then east into the woods. Petey was whining with eagerness for the first three hundred yards down the slope then milled around in one spot. As the shadows lengthened through the woods Orvil and Petey cut larger and larger circles around the dead point but it was obvious to Orvil that the boys trail was gone. He radioed in.

* * * * *

"Commander sir!"

The Colonel reached over to his bed and touched his shoulder mike. "Yes?"

"I'm getting many strange messages from the local police, Sir…"

"What kind of messages?"

Lieutenant Pong-Uk reached over the shoulder of the watch radioman and tweaked some dials. "Sir, apparently two hikers are twenty four hours overdue to the south of us and they have search parties out all around our area—but nothing has been found."

In his private apartment cum office Colonel Pong-Yul sat in the Sukhasana (sit/easy) position on a reed mat on the floor with his writing pad on his right knee. It was a matter of great personal pride that he could still assume the difficult position and concentrate. He had been bringing his personal log up to date. He drew a deep breath and closed the book. "Dam! G-101! Radio the squads immediately. There is to be no contact with the searchers on pain of death. They are to go to ground and stay there until this clears up. I repeat, there is to be no contact with the local police force or any other search party, is that understood?"

"Yes Colonel Sir."

* * * * *

As dusk slid into night Sheriff Gibbs' phone rang. He grabbed it off the cradle. "Yes."

"Wayne, Charles Burlingame here."

"They's nothing new, Mr. Charles. Petey lost the boys trail between 1 and 2 and the rest of the deputies found nothin."

"I know," Burlingame said heavily. "You'd have called me already if there was. Looks like we've got something pretty serious here, doesn't it?"

"Fraid so."

"You're boys have done a wonderful job Wayne, nothing against them, but I think you aught to call in the GBI[1]..."

Sheriff Gibbs smiled, it was nice to be half a step ahead for a change. "I talked to the Commander down to Atlanter this afternoon, Mr. Charles. He just called me back a few minutes ago. They're goin to lend me a chopper tomorrow—it'll be here by first light..."

Relief was evident in Burlingame's voice. "God, that's great Wayne. I'm really worried about the boys."

"Well, Ah'll over fly the Devil's Play Pen, the caldera and the rocky area up around the crevasse. Them area's are nigh impassable on foot and we ain't been able to do justice to 'em. But Ah can cover a lot of ground in a day with the bird and Ah should be able to spot anything unusual."

"How big's that chopper?"

"Jist a two seater."

"Oh." the disappointment was heavy in Burlingame's voice.

"Ah'll be the observer, Mr. Charles and I'll keep yew posted by radio if we see anything—anything at all."

"Wayne, I appreciate that. The Chapmans, that's Harlowe Chapman's mother and father are coming into Atlanta tonight from Phoenix, Arizona. They'll drive up tomorrow and will be my house guests until this mess is solved."

* * * * *

Twenty minutes later the Sheriff's phone rang again. He lifted it impatiently, the way it was ringing he knew it was from out of town. "Sheriff Gibbs here."

The voice on the other end was crisp and authoritative. "Sheriff Gibbs, Leon Thurmond here. I work for the Atlanta Gazette and I'm also a stringer for United Press News. I understand you've had some weird disappearances up there recently and you've got a police chopper arriving there manana?"

Gibbs smiled grimly. "Funny thing 'bout bad news, it travels at the speed of light. Yes, We've got two boys missing and we intend to find…"

Thurmond cut in. "Is it true that one of them is the son of the richest man in the county, Charles Burlingame?"

"That's right."

"Thanks, I'll see you in the morning." The line went dead before Gibbs had a chance to answer.

CHAPTER 9

The Georgia Bureau of Investigation was true to their word. A two seater lifted off the police pad at the North East Precinct of the Atlanta Police Department at 04:30 hours and dropped onto the chopper pad at the Caravelle County Hospital thirty five minutes later. Gibbs met it and was briefing the pilot when a white TV Van pulled into the parking lot. A lanky figure unwound from the passenger side and approached them. He appraised Gibbs uniform.

"Sheriff Gibbs?"

"Yes."

He offered his hand to Gibbs. "I'm Thurmond from Atlanta."

Gibbs said, "Glad to know yew. This is Lieutenant Gabralski, Chopper pilot with the Atlanta Police Department."

Thurmond smiled. "Yes, Sheriff, I know the Lieutenant." He shook hands with the pilot. "Gabby and I go way back."

Gibbs flipped his thumb at the truck. "Did yew have to git TV mixed up in this?"

Thurmond shrugged. "Everything is news these days, Sheriff, especially when it concerns the rich and powerful. But aside from that we have scanners and cellular phones in there so we can monitor all the police bands. You can consider this your portable electronics center for the day."

"Ah wonder how we've ever made it before now without electronics."

Thurmond cocked his head sideways, "While you're in the air I'm going to talk to Burlingame but we'll be listening. Good luck in your search."

Gibbs thanked him. Within five minutes he was strapped into the passenger seat and they lifted off into a velvet sky rapidly turning slate gray behind fading stars. As they lifted above the dark line of the treetops Sheriff Gibbs swallowed swiftly to stave off the budding nausea.

*****

"Sir, there's a chopper quartering The Devil's Play Pen."

Pong-Yul looked up from his reports. "Who is reporting?"

"Squad Number three, Sir."

40

"Where are they?"

"In the woods just south of the Play Pen."

"How big is the chopper?"

The Watch Commander spoke briefly into his mike, listened intently than turned to Pong-Yul. "A two-seater Sir. a Robinson R22 Beta. It bears the markings of the Atlanta Police Department."

"Are you monitoring their band?"

"Affirmative."

"They will eventually want to land here. Normal procedure dictates that they request permission unless they have a life threatening emergency. Let me know the moment they hail you."

"Yes Colonel, Sir."

* * * * *

As they swept through the sky the first half hour was very uncomfortable for Sheriff Gibbs. A ground person by nature, his inner ear did not easily tolerate the gliding flight of the tiny chopper. But as soon as the ground began to come out of the shadows he wrapped himself in the search and forgot he was in the air. It was the first time he had searched from a chopper and he found it was like suspending yourself over a table laid out with miniature trees, lakes, houses and mountains. Eventually you forget you are aloft and you concentrate on the search. Everything below you looks like an architect's next project.

An area devoid of trees just south of Voodoo Mountain and north northwest of the lake about one and a half miles east to west and a mile north and south, The Devil's Play Pen was a scarred, pockmarked moonscape in the early morning light. Gutted with sinkholes and caves, small fumaroles belched hot air and smelly gas into the warm sunlight creating spooky little fog banks and strips of mist winding through the eerie landscape. Animals avoided it because there was no water and the sulfur and rotten egg odor offended even them. Usually deer went around it, but occasionally they crossed the area to avoid predators on moonlit nights. After a good long look, out of town hikers normally skirted the area and local hikers rarely ever entered it. A virtual no-man's land was what Gibbs saw as the chopper drew imaginary grid lines back and forth. Nothing moved below and within twenty minutes the Sheriff was satisfied and asked the pilot to move up into the caldera.

The pilot swung the bird up and over the edge of the caldera and down into the maw of the volcano. The green, shaggy crags rose above them as they slid into the perennial half-gloom. It was a drop of nearly two thousand feet to the valley floor and the pilot did it cautiously like a slow leaf swirling in the autumn wind.

* * * * *

"The chopper is coming into the valley, Colonel Sir."

"Is the rocket crew standing by."

"Yes sir."

"Good. Have them ready to fire if I give the order. When the chopper hails I will answer."

"Yes, sir."

* * * * * *

By the time they reached a hover some two hundred feet above the antenna mast Gabralski had given Sheriff Gibbs a look at the whole valley floor, including the layout of the three small buildings and the fenced-in compound.

He punched buttons on the console and indicated Gibbs should talk to the ground.

Gibbs was matter-of-fact. "Voodoo Mountain, this is the police helicopter hovering above yew, are you listenin?"

For a few moments there was no acknowledgment from the ground but as he was about to repeat his question, the radio silence was broken by a single, sibilant word, "Yes."

"Sheriff Gibbs here, of the Caravelle County Sheriffs' Office. We have two young male Caucasian hikers lost in the woods here abouts and Ah'd like to know iffen yew know anything about them?"

Colonel Pong-Yul measured his words carefully. "Sheriff Gibbs, I am Doctor Sul of Ident Inc. We have no personnel above ground here. Our security has not been breached and our fence is secure so I cannot help you."

Gibbs looked at the fifty foot antenna below him. He hated long range conversations, he liked a man in front of him so he could see his face. "Thank Yew Doctor, I was hopin yew could hep us out. D'yew mind telling me what yew folks do down there?"

Pong-Yul's voice was sincere, ingratiating. "We do advanced cancer research, Sheriff, something we hope will benefit all mankind."

"Wal, keep up the good work Doctor. Iffen yew should hear anything of our boys I'd appreciate it if you'd call me on your radio and let me know."

He flipped the radio switch and signaled the pilot to lift out of the caldera.

* * * * *

"Where to now, Sheriff?"

Gibbs pointed to the horizon. "Bear west and a little south. I want to take a look into the crevasse."

"Crevasse?" Grabalski asked, arching his eyebrows, "What's that?"

"It's a natural split in the rock. About a hunderd and fifty feet long, a hunderd feet deep, only ten feet wide and closed at the south end. Bottom is almost level. The north end opens into a series of small caves."

"Anyplace to land?"

"No. The west side of it is forty feet higher than the east and the ground slopes real bad on both sides. Nearest place to land is about a half mile south."

The pilot was watching the trees flow under them. "How we going to see into it?"

"Midday like now when the sun is sittin straight up is the only time the bottom is visible and then none to good. Your searchlight may hep—we'll see. There it is over there." He pointed down in front of them. "See what looks like a cliff stickin up. Over there. Pull in above the cliff but hover on the east side so I can look down in. Be careful, the wind gits kinda strong comin out the south end."

Indeed as he spoke an up-draft from the southern end of the crevasse pasted Gibbs hard against his seat and wafted the little chopper five hundred feet straight up into the clear air. The pilot grinned, swung the bird around in a graceful arc and jockeyed it back down to hang just outside the shaft of air.

"I"m going to tip it a bit to your side so's you can see better."

Gibbs nodded and caught his breath as the chopper tipped on its side with the pilot above him and he was sitting there seemingly suspended in mid-air hanging above the crevasse.

He directed the pilot to move slowly along the lip of the massive gash. Even though a strong sun sat straight over head the bottom of the split was hooded in shadows. But there was enough light to show something there in the bottom. He held up his hand to stop the movement of the chopper, steadied himself and focused his binoculars. He distinctly saw a leg and it ended in a hiking boot.

They skimmed farther along the lip of the crevasse and he made out a hiker's pack and what appeared to be a second body hidden in the deep shadow.

How in the world could these two experienced hikers have ended up here. It stunk of foul play, he thought, but why, and who. Who would want to hurt these two kids up here?

They both had to be dead. Nobody could fall down or be thrown down that precipice and survive. He motioned the pilot to return to base and flipped the radio on.

CHAPTER 10

"Colonel Pong-Yul Sir."

"Yes?"

"They have found the American hiker's bodies."

"Where?"

"Somebody dumped them in the crevasse."

"G-101?"

"No word on that Sir."

"Has to be him. Are you monitoring their radio direct or are you talking to one of our squads?"

"Squad 3 is just six hundred meters northeast of the crevasse, Sir but I am hearing the helicopter direct."

"Remind Squad 3 to stay out of sight of that chopper. I'll be down in a few moments to listen to that tape."

"Yes, Sir."

CHAPTER 11

It rained long and hard until the early morning hours.

The salvage party left the Convenience Store just after the rain stopped at 4 a.m. and picked their way up Trail 1. The woods resounded to the putt-putt engines of four wheelers and three small donkeys.[2] As the morning warmed, mist swirled around the searchers and turned the open patches of the trail into foggy daylight.

* * * * *

Harlowe Chapman Sr and Charles Burlingame had demanded the right to go with the rescue party but Sheriff Gibbs had convinced them this was strictly a salvage operation. The day would be long, hard, full of difficult climbs and it already had a gruesome ending. It was no place for someone so emotionally involved. They finally saw the reasoning and elected to wait beside the radio for news.

Leon Thurmond appeared in rough clothes and hiking boots and claimed a spot on one of the donkeys. Gibbs was less than overjoyed but nodded his head and the caravan crawled off into the woods up Trail #1.

Sunup was a dreary affair with gray, scudding clouds chasing patches of blue across the mountains. The trail was mushy and designed only for foot traffic and it was mid-morning before they reached the point on Trail 1 where it would be easiest to move up to the crevasse.

Even with the advantage of the tiny four wheelers they could not reach the gash in the rock. They parked two hundred yards southwest and below it and prepared for the nearly vertical climb.

Gibbs dispatched Deputy Orvil Vaughn and Old Petey immediately so they could search both sides for scent before the main party reached the crevasse.

Orvil started Petey at the south end on the east side and they slowly worked their way north. Petey whined, wagged his tail furiously and snuffed at everything but the lip of the crevasse was rock and the rain had washed any scent away. Back twenty five feet from the rock shelf the dirt was disturbed but the disturbance resembled nothing human. Orvil pushed ribboned stakes into the ground around the periphery of the disturbed area so the party would not trample through it.

After two arduous hours on foot lugging equipment up the steep, rocky hill the Sheriff and the main body of searchers arrived at the lip. Orvil indicated the staked area and proceeded around the hill to go up on the west side of the crevasse and let Petey finish his job. Thurmond circulated with a portable Video Camera and began recording.

*   *   *   *   *   *

Gibbs directed the set-up on the lip as two of his deputies readied themselves for the descent. A small gas generator was cranked and floodlights were lowered into the crevasse. The remains of two bodies, surrounded by the debris of their packs, became instantly visible. They were lying within fifteen feet of each other on the cold, sloping, boulder strewn floor.

Gibbs went over to one of the deputies who was suited up for the rappel into the crevasse. "Billy, I want yew down there first by yorsef. Set up yore lights and take pitchers from one end of that hole to the other. I want Polaroids of the bodies frum four different directions and anythin else interestin. Number them pitchers as yew take 'em and when yore through yew send 'em all back up to me in the basket, then Ah'll send Otis down to hep, ya'hear."

"Gotcha Sheriff."

"Mind yew, don't step on nothin or move nothin until yew've shot it. There may be some bad smell down there—yew got your mask and the Benzoin?"

Billy patted his vest. "Yes sir."

"Git."

A squat, aluminum tripod four feet high had been anchored in the dirt twenty feet back from the lip. Billy attached his rope to this, adjusted his rappelling harness, walked to the edge, threw the coil of rope into the void, turned around, attached his descender, winked at his partner Otis and walked backwards over the precipice.

As Billy disappeared below the edge Gibbs felt that old familiar tightness between his ears and he swallowed trying to relieve it. He had sent Billy Matthews and Otis Creedmont to Climbing School at the county's expense and had gone to their graduation examination.

That was the first time he had noticed the tenseness and it happened each time he watched them drop over.

He fanned the rest of the deputies out in the area to look for any evidence they could find. Then he sat down by the edge of the crevasse with a walkie-talkie and waited for Billy to talk to him. Thurmond waited silently near him.

* * * * *

"Sheriff Gibbs."

"Yeah, Billy."

"They's two bodies down here all right. Been here at most a couple days. One's blond, one dark."

"Yew recognize either one a them?"

"Sheriff they's a mess. The critters been workin on 'em pretty good. Their faces, eyes 'n ears and most a their fingers been et off but the blond one has got to be Benjie. The dark one must be that Chapman boy."

"Anythin unusual besides that?"

"Yeah," came Billy's hollow voice. "No blood."

"None at all?"

"I ain't moved anythin Sheriff, but it looks like what little they is is right under their heads. Ah spect they was dead when they was dumped in here. Both got broken arms and legs."

"Whut kinda junk is layin around the bodies? Anythin odd about that?"

"Yeah, they ain't no water nor food, no knives or other gear. The animals coulda et the food but not the water. Nobody comes up in here without bringin water. Nobody comes up here without weapons of some kind."

"Take yore pitchers and Ah'll send Otis down."

CHAPTER 12

"Colonel, Sir. The Americans have recovered the bodies and they are leaving the woods."

"Bring the two northern squads in. Tell the two southern squads to intensify their efforts, we need G-101 back before he does something else stupid."

* * * * *

"Sheriff, let's cut through all the rest of this crap. My question is when are you going to release my boy's body?"

Sheriff Gibbs turned to the speaker and shrugged with his hands. "Ah'm not in charge of that, Mr. Chapman. We don't have the resources here to do an honest-to-god forensics autopsy so the GBI does it for us. Once yew call them in they work on their own timetable."

Gibbs, Chapman and Charles Burlingame were seated in Sheriff Gibbs office. He had given both of them as much information as he had but had refused to let them see the bodies prior to shipment to

Atlanta. What was the point when the salient features on both bodies were missing. Now it was a matter of fingerprints and teeth since neither boy had any tattoos or birthmarks. Both parents had supplied him with dental records and their schools had faxed the GBI copies of their thumb prints.

His phone rang, he answered it and listened gravely for some three minutes while his companions were visibly chaffed.

He put the phone down. "That was the GBI. They have positively identified both boys. They agree with us that they were dead before they ended up in the crevasse…"

"Somebody threw them in!" Chapman said furiously.

"Yes." said Gibbs.

"Wayne, did they say how the boys died?"

"No, they can't find out…"

"Damn…"

"They are requesting your permission to send both boys to the FBI in Washington."

"For God sakes, what for?" Chapman queried

"The FBI has the best forensics team in maybe the whole wide world. If anybody can tell us what happened to them they can."

"I'm for it," Burlingame said.

"I don't know, Mrs. Chapman won't…"

"They need both boys for a positive conclusion." Gibbs confided.

"Do it Harlow, it's the only way to find out."

"All right Charles. Jr's mother will kill me but I guess there's no other way."

Gibbs picked up the phone, dialed Atlanta and gave them permission to ship the remains to Washington, DC. while the bereaved fathers listened.

"He said it would take probably ten days total to find out. Are you going to stay in town Mr. Chapman?"

"No I can't stay away for that long, and there's no reason to now. Can you make sure they send Jr. to me in Phoenix when they're done?"

Burlingame stood up. "Sheriff Gibbs and I'll see to that, Harlowe. Now, let's go tell Mrs. Chapman where matters stand."

Sheriff Gibbs watched the two walk out together. He saw two men now bound together forever through a common bond which neither would ever have willingly inflicted upon the other.

CHAPTER 13

TEN DAYS LATER

"Sheriff, Atlanter's on the phone."

Gibbs snatched the phone to his ear, "Sheriff Gibbs here."

He listened intently for nearly five minutes while uttering only sporadic agreements. He also made several notes.

He put the phone down, called all the deputies to his office and dialed Charles Burlingame.

"Mr. Charles, Wayne Gibbs. I just got a call from the GBI in Atlanta regarding the boys."

He paused, chewing on his lip. "No, not the FBI, this comes from the Smithsonian." He consulted his notes. "Actually the word comes from The Curator of The Department of Anthropology, National Museum of Natural History at the Smithsonian Institution in DC." He paused. "Yes, Sir. The FBI consulted them. Here's what they said. They said both boys was murdered and both was killed by a crushing upward blow to the neck." He stopped for a moment and the phone crackled.

"Yes, they was dead before they went into the crevasse. It was the pathologist's opinion that the blows were decisive and they both came from the same man since the blows was identical. They also said they think we are looking for a very short, strong male and that he most likely is extremely proficient in the martial arts."

CHAPTER 14

Billy Matthews skidded to a halt in front of the Sheriff's Office, jumped out of the patrol car and raced into the building.

Gibbs watched him run the length of the office waving a newspaper. Billy burst into Gibbs' office.

"Sheriff, looky here, Atlanter's got a big story. Look what they're callin our murderer."

Gibbs picked the paper out of Billy's hand. "Billy yew in pretty good shape but yew still about to have a hart attack." He glanced at the paper.

The two inch headlines screamed out at him. "WOODCHOPPER KILLS HIKERS!" He read the second headline. "Unidentified killer stalks NE Georgia woods, kills with bare hands." The lead story began, "About two weeks ago two young hikers disappeared in the mountains in Caravelle County. Ten days ago their bodies were recovered from a deep crevasse and today we learned from the FBI that both were murdered by the same method, a violent chop to the throat…"

CHAPTER 15

The Woodchopper stood within the edge of the woods at the rear of the convenience store just outside the ring of light thrown by the pole lamp. He had come exploring over the ridge and through the woods after watching the Sheriff and his men raise the two bodies he had left in the crevasse. He did not know, nor would he have cared, but he had crossed the line and he was in Cherokee County. The food he had gleaned from the boys packs had run out twenty four hours ago. He knew where there was light there was food. He was hungry and thirsty and in pain.

He circled to the right within the tree line until he could get a look at the front. Two employees had just locked up and were walking toward their cars in the parking lot. A large, young man and a young woman with flowing hair. He waited silently. The young man said something to flowing hair than leaned across the hood of the nearest car and engaged her in conversation. The large young man's back was to the store.

The Woodchopper slid back through the trees and crept across the yard to the rear door. A padlock secured it. He wrapped his blunt, powerful fingers around the lock and threw his shoulders into the twist. To his delight it groaned and the hasp of the lock broke open. He slipped inside.

He was at the end of a long hallway very dimly lit by one small bulb at the far end above an open door. He walked quickly toward the bulb and paused. He looked through the door into a larger room stacked high with boxes. This stockroom in turn lead to the front of the store which was brightly lit. He tip-toed across the stock room and peeped carefully around the door.

He was concealed from the front of the store so he slipped out, reached into a cold case, crammed half a sandwich into his mouth and washed it down with a cold bottle of cola. As he ate he continued to pick up items from the shelves and stack them near the back door.

* * * * *

The large young man leaned against the hood of the car and said conversationally, "We could have a lot of fun at the lake on Saturday."

She stopped as she was about to unlock her car door. She turned to him, He was seventeen, a year younger than she and such a child. Be gentle she thought. "Sammy, I can't go out with you on Saturday, I've got other plans."

His face colored but he laughed easily. "Is his name Tom, Dick or Harry?"

"No, its not another boy…" He winced and she was immediately sorry she had used the word. "I'm not going with someone else, but the folks have requested I go with them to Atlanta and…" suddenly her attention was attracted to the store. She was positive she had seen a subtle shadow through the window. "Sammy…" She stopped in mid-sentence.

"What's the matter?"

She kept her eyes focused on the front of the building. "I could swear I saw something move in the store."

He turned around. "I doubt it, shucks, we just closed up and I know there was nobody inside…" He stopped as the light level changed slightly in the store, "but there is something funny with the lights. I'll go back and check it out."

She walked with him quickly back to the door and he pushed his key into the lock. The big steel bolt turned with a clunk and he jerked the door open. They stood there briefly their hearts pounding in their ears and heard startled footsteps in the back of the store.

"There is someone here," Sammy yelled. "I'll get him."

"No Sammy, let's call the Sheriff."

Sammy was already running down the aisle. "It'll take the Sheriff fifteen minutes to get here.

"I'll go with you."

"No, he's going through the back. Run outside 'round back, try to get a look at him."

Sammy broke through the back door at almost the same instant she rounded the back corner of the store. They both saw a squat childlike figure, with two plastic bags of groceries, running rapidly toward the darkness of the woods. They followed him to the edge of the light, then very wisely turned back and called the Sheriff.

* * * * * *

"Commander."

"Yes?"

"There is an 'FYEO'[3] coming in Colonel, Sir."

"Bring it to me."

"Yes sir."

At the knock on his door he said "Come" and an orderly, back stiff with respect handed him a closed packet. The Colonel dismissed him with a wave of his hand and pulled his cipher book from the back of his desk. As he decoded the message a cold wave passed over him from head to foot. It was a message from the Supreme Commander of the Military Affairs Commission of the Central Committee, General Kye Yong-Sop. The two most important things to remember about General Yong-Sop was that, first he was the President's cousin and second, he was a hard, brilliant man, totally without pity. Pong-Yul, having idolized him for more than two decades, had spent long hours studying Yong-Sop's background and could recite the litany of the General's deeds from memory. His General's words were brief and brutally to the point.

"You are engaging the enemy far more than is considered necessary or desirable. The major purpose of this project is close at hand and must not be jeopardized. You must contain yourselves until that time and not alert the Americans. Caution your men that no infractions will be tolerated. Any transgressors are to be summarily executed. End"

CHAPTER 16

The Curator at the Smithsonian Institution graciously offered to cremate the boys remains and did so after receiving permission from the two Chapmans and Charles Burlingame. The cremains were then shipped to the respective parents accompanied by a note of condolence from the Curator himself. Burlingame transferred Benjie's remains from the plastic box he had come home in to a simple pewter urn. Neither he nor Benjie were ready for funeral services yet so he sat the urn on the mantle above the fireplace and surrounded it with the cards he had received from friends and well-wishers. One in particular attracted his attention. It was from his old friend, reclusive genius Jason Amador. The feminine hand meant that Amador's long-time assistant Christa Bonamay had addressed it for him. No matter, it was the thought that counted. He propped it right next to Benjie.

* * * * *

Burlingame went back to his office and the family businesses with an uneasy feeling. There were a great number of unanswered questions in his mind. He felt like a man waiting for the other shoe to drop. Who did the killing? Why were the boys killed? Where were they killed? Why couldn't they find the killer?

He posted a ten thousand dollar reward for information leading to the arrest of The Woodchopper and decided that as soon as he could sit down and talk about the double murder rationally he would call on his old friend, Jason Amador.

CHAPTER 17

The pains were like labor pains now, coming more frequently. They wiped him out and when he came out of the red fog he was ravenous. The two hikers netted some food but it had run out quickly. The raid on the store replenished his stores for a few days but with his appetite the food was fast disappearing. When he left the store with his arms full of groceries he had run back up the ridge and over the other side before dropping into the soft pine needles, gasping for breath. His great strength was still with him but like a Cheetah in hot pursuit he over heated in a very short distance and had to stop to cool down. He laughed, just a mirthless little cackle, as he sorted through the items. This time he had thought to snag some aspirin. He downed four of the pills hoping for some relief from the horrible cramps. He crawled into a copse and stayed there for two days—eating and sleeping, fighting through the spasms of pain.

On the third night G-101 ventured out and dead reckoned his way east and south and ran across Trail #1. He knew anyone looking for him would be staying on the trail so he shuffled along on parallel and continued south, dropping into exhausted sleep when the waking world became too painful to endure and waking when the enduring bouts of pain forced him back to the surface. He catnapped most of the daylight hours and stumbled fitfully through the night. The dark and the pain led him along erratic paths but the downhill slope drew him inexorably towards Highway 115. On the fifth night as the moon lifted above the edge of the deep blue sky he came to the dark, calm waters of Flint Creek. He dropped at the edge, plunged his face below the surface, flushed his nose and mouth then laid back and focused on the stars peeping through the trees. His mind wandered from this topsy-turvy world, far from home and full of pain. He came out of the shadows of the jungle into a small clearing, walking towards a little hut basking in the steamy sunlight. A tiny woman with cherub face paused in the door, saw him—ran to him, arms outstretched, her mouth twisting with yelps of joy. He opened his arms, grasping, clasping, to gather her to his chest but roils of pain pushed her back— back—back into a red fog. He rolled over, biting into the pine needles, grunting and straining with the pain and submitted to the fog.

* * * * * *

The moon was well beyond center when The Woodchopper opened his eyes and felt the continuing throb in his gut. He sat up, painfully and picked up his final can of baked beans. He opened them with his knife and began scooping the cold goobers out with his fingers.

As he ate he heard music in his head. Or at least it seemed so. He licked the final sweet juice from his fingers and laid back with closed eyes—then sat upright with a start. He *was* hearing music. Very faint, but melodious and rich it coursed through the woods and assailed his ears. Just above the threshold of his hearing. It must be a long way off.

He got to his feet and slipped down hill through the woods, a child seeking the Pied Piper. The notes grew louder and soon, standing on a small knoll which gave him a view of the down ward slope, he saw an aura of light above the woods. He resumed his downward trek and presently, a hundred yards below him through the trees, he could see a road and on the other side a brightly lit convenience store. From in front of the store came the music.

He moved down the slope and stopped in the final fringe of brush and trees just before the road. He dropped at the base of a tree and studied the scene. Someone was sitting in a car in the front parking lot and they were listening to loud, loud music. It flailed his ears, and boomed and reverberated through the woods. No other sound could be heard as the decibels blasted out from the open windows. It was so loud it hurt his head. The WoodChopper put his hands on his ears and started across the road.

CHAPTER 18

"O my Jesus, Gawd. Oh Jesus, Oh my Gawd."

"Billy, whatsa mattah with yew?" Deputy Sheriff Otis Creedmont yelled as he was coming around from the far side of the patrol car in the parking lot of the convenience store. Billy had just gone up the three steps and entered the building. The horror in Billy's voice was enough to make Otis draw his gun and run into the store in a crouched position. He came up behind Billy standing in the aisle. Billy was so shocked he still had his pistol in holster. Otis looked around Billy and sucked in his breath. The floor five feet in front of Billy was covered with a monstrous blob of blood. It glistened in the lights and there was enough of it to raise up at the edges and look like crimson plastic on the floor. From the other end there were drag marks along the aisle and it was obvious to both that someone had been butchered and then dragged into the store room.

Otis tapped Billy on the shoulder and whispered, "We gotta look in there." and he pointed to the store room door. Billy nodded numbly and motioned Otis down the side aisle and around the end so he could

74

check the rest of the store at the same time. He began to skirt the blood and the two began a crab like scrabble toward the store room door.

The lower corner of the store room door was brushed with blood. Billy inspected the knob, found no blood and turned the knob with two fingers. Just as Otis joined him the door gave easily and swung inward. The light was on inside the store room and seemed to push the rank odor of urine and feces out and around them.

Both gasped. There was a small body lying face down on the store room floor, the feet just clear of the swing of the door. The body was heavily splattered with blood.

Billy stepped gingerly into the room and reached down to check for signs of life. He caught his breath. It was the body of seventeen year-old Kathleen Burlingame, Charles Burlingame's youngest child and only daughter. Her head was turned to the left against the floor. He could not find a spot not covered with blood so ignored it and touched her high up under the jawbone looking for a pulse. Finding none he tapped her gently on the left eye desperately seeking for some reaction. There was none.

He turned to Otis, large tears brimming his eyes. "She's dead Otis,—she's dead. This little girl was like mah kid sister an she's dead Otis, somebody just butchered her."

Otis laid his hand on his partner's shoulder. "We'll find out who did it, Billy. Right now we got to find Lester. He's supposed to be on this shift with Kathleen."

* * * * *

Can't move very fast, he thought. These plastic bags and the backpack make it easier to haul food but still it is difficult to move quickly. Have enough food now for two weeks or more. Tired and want to rest. Must keep moving. Soon there will be dogs. When I get to the stream will walk for a while in the water then out on the bank and spray my shoes with bug repellent. A simple way to confuse a simple organism—a dog.

CHAPTER 19

"Commander!"

"Yes?"

"The American's radio is broadcasting another double murder."

"I will listen to the tape in a moment. Does it appear to be G-101?"

Lieutenant Pong-Uk paused. One must be careful with any Colonel but especially this one. "If the Americans are telling the truth it has to be him, Colonel, Sir."

"Put out two more squads to the south. Give them each a quadrant of the southern section. Same orders as before only it is more urgent now than ever that we catch G-101."

"Yes, sir

* * * * *

"We don't know everything yet but here's what we do know." The speaker was Alfred Parmentier, Special Agent for the GBI, Atlanta Office. A squat, balding man with perennially loosened tie, he

77

exuded pugnacious authority. It developed that he was a very methodical, tenacious investigator.

It was three days after the deputies had discovered the carnage at the store. Parmentier had one leg astride the corner of Sheriff Gibbs desk. He was talking to Sheriff Gibbs, Charles Burlingame, Cecil Westmore; Lester Westmore's father (Lester was the other clerk at the all-night Convenience Store), and Leon Thurmond who had become a fixture in town since the convenience store murders. Also present was young Turley Crenshaw, cub reporter from the Greenville Sentinel on his first foray afield. Turley was a large, impatient young man with a wrestler's neck and a sweaty, receding hairline.

Sheriff Wayne Gibbs had taken one look at the Convenience Store Crime Scene, taped it off and had called the GBI. Beyond that he had put Old Petey and Orvil on the scent but they had entered the woods and lost all direction when they came to Flint Creek.

Parmentier's voice softened as he looked at Burlingame and Westmore. "Both youngsters were killed with the same knife. Looks like one of those survival knives with a leather handle and a serrated back edge…"

Burlingame took a deep breath. Parmentier stopped and looked at him with raised eyebrows. He asked, "Was that by any chance the same kind of knife your son carried into the mountains when he was killed?"

Burlingame nodded.

"The WoodChopper," Crenshaw said under his breath.

"Yes, it looks like it," Parmentier agreed. "So we probably have the same perp. I hate to say these things to you two fathers because I know they're painful as hell but I've got no choice."

Burlingame waved him on, Westmore snapped, "I wanta get that son of a bitch, get on with it!"

"Whoever did this evidently went to the car first. Your son must have been sitting there during a slack period listening to his tapes. The killer just reached through the window, grabbed him by the hair, snapped his head back and slit his throat. Then he pushed the body across the seat and turned the radio off."

Parmentier cleared his throat. Even after twenty five years in this job he had never gotten used to this stinking part. "He then went into the store and ran into the girl right in the middle of the aisle. She must have been shocked into inaction because she stood there while he stabbed her three times…"

"You said she was stabbed twelve times."

"Yes, Mr. Burlingame she was, but the ME[4] says that she must have turned to run and the killer jumped on her back and the other nine times she was stabbed were either on the way down or on the floor."

Charles Burlingame closed his eyes in a futile effort to stop the tears. He was sobbing quietly as Parmentier continued gently. "She wasn't sexually assaulted, Mr. Burlingame, we can thank the Lord for that. He then dragged her into the store room and was damned careful to not leave us a footprint. We have some smudgy fingerprints but the FBI says they are too vague to get a make. The way it looks after he finished with the girl he went out into the store and started munching on some raw vegetables while he put together a knapsack of food…"

Cecil Westmore broke in." How 'bout that picture she drew on the floor in her own blood?"

Parmentier sighed and produced a Polaroid shot from the crime scene. It showed a drawing on the floor—a drawing made in bright blood with a small finger. It was of a perfectly round head, two dots for the eyes, two dots for the nostrils, a down-curving split for a mouth, a round, bald head, no eyebrows, no lips on the mouth. He said, "Every person we have in the Department has eyeballed this stick drawing and we've all come to the same conclusion. It's a crude drawing made by a young, non-artistic clerk who was dying in great agony. It gives us absolutely nothing to go on."

Burlingame blew his nose and reentered the conversation. "How about that list of everything missing?"

Parmentier consulted a typewritten sheet. "We agree with you that with the exception of the food, water and Saran Wrap all the missing items are those kinds of things that would be ordinarily shoplifted during the season. However," He shifted the sheet to his left hand and pointed to it with his right. "clothesline, water, flashlight and batteries, can opener, small pulleys, waterproof matches, mosquito repellent, aspirin, a knapsack... These are all items that would also be needed by someone who is living off the land. We think he's living up here in your hills."

Burlingame looked in turn at Sheriff Gibbs and Westmore. "Since we gave you that list, Mr. Parmentier, we discovered that the wall map of the Voodoo Mountain Area is also missing from the store. Now he knows the whole area. That means we can expect him to hit again."

Parmentier was matter of fact. "As soon as he gets hungry, Yes."

Gibbs asked Parmentier. "How much help can you give us?"

"We're an investigative organization Sheriff and Georgia has one hundred and fifty eight counties. We have no personnel to leave on the premises, I'm sorry but you'll have to mobilize locally." He began putting his papers back into a tan leather briefcase. He turned to Burlingame. "Thanks for the tip on your daughter's red and black nylon jacket being missing. We had a report filter through of a convenience store looting over in Cherokee County about ten days ago. The kids had just locked up and somebody entered the store from the rear. The kids almost caught him. They saw him running into the woods at the rear. Was a little guy wearing pants and a shirt, no jacket. Now I think he has a jacket. Considering what we saw here it's a damned good thing they didn't catch him."

CHAPTER 20

"Sheriff, Ah'm gettin a fax in from the GBI in Atlanter."

Gibbs looked up from his desk where he had been idly poring over the theft list from the convenience store. "What's it on, Billy?"

"Looks like an answer on the fingerprints from the convenience store."

"Read it to me."

"Says they got no match. However they determined that natural body oils were present in the smudges indicating that the suspect was not wearing gloves or, Jesus!…"

"Whatsa matter, Billy?"

"Boy these guys don't pull no punches. They say maybe he was wearing gloves and had masturbated with the gloves on. You don't suppose that pervert…?"

Gibbs chuckled. "No, there was no evidence of anythin sexual around Kathleen's body—just that pee…"

* * * * * *

"Have the squads anything to report on G-101, Lieutenant?"

"No Sir."

"What's the last report we have on him?"

Pong-Uk sighed but not loud enough to be over heard by the Colonel. He had been asked this same question a dozen times. "Over a week now sir."

"Have the squads made any contact with the search parties?"

"No sir, they have observed them numerous times but they are following your orders to the letter."

"Good."

* * * * * *

Charles Burlingame stepped back from the mantle and looked up at his family. Benjie and Kathleen rested side by side in identical urns surrounded by cards from almost everyone in the community. Kathleen's horrible death coming so close on Benjie's left him a hollow shell of burning grief with no tolerance for a large funeral. He had had a simple memorial service for the two of them, by invitation only but the whole town turned out to share his sorrow. He barely heard the eulogy spoken by Pastor Malone who had known the kids since he had baptized them. Burlingame stood with folded hands, gazing numbly at the pewter urns as friends and well wishers brushed by to pay their final respects. Lost in his anguish he did not offer to shake hands with each of them but received their shared grief and compassion as they placed a gentle hand on his shoulder or brushed against his elbow. This is all just a bad dream, he told himself, and I'll wake up. Kathleen will come home from school, throw her books on the hall table, grab a coke and run upstairs to argue with Benjie.

The simple pewter urns gleamed dully in the light of the chandelier.

* * * * *

Lake Pawnoo was shaped like a skinny football with its north end lying north northeast. Roughly four miles long by one mile wide and in places up to five hundred feet deep it had been formed during the Devonian Period three hundred seventy five million years ago when the Blue Ridge became the southern spine of the Appalachians. It was already an old body of water when Voodoo Mountain blew it's southeast face into the antediluvian sun and created the caldera and The Devil's Play Pen.

Armed with his newly acquired map The Woodchopper made his way up along the north eastern side of the lake until he rounded the tip and stood in the woods looking at the northernmost cabin of the twelve that ringed the southeastern side. This cabin was at least two hundred yards from its nearest neighbor to the south and secluded in the trees. It was clad in brown wood and topped with wooden shakes. The front facing the road was punctuated with a large picture window alongside a fireplace. On the side wooden steps led up to a deck half as large as the cabin. The deck stretched around the side of the cabin giving elevated access to the back door where another set of steps led back to the ground. The back deck was large enough to reach out over the edge of the lake.

He watched for two hours through the last rays of the sun and into the beginning of the long night. There was no movement and no lights came on as darkness fell. Finally satisfied, he slipped through the shadows and stepped lightly up the back steps. The back door resisted his urging only momentarily and he slid breathlessly into a dark kitchen lighted by moonlight through a single window. Sharpened senses told him he was alone. The refrigerator door was blocked open just a crack. No light showed. He ran his hand inside and it was room temperature. This family was not expecting to be back for awhile.

He turned his attention to the rest of the cabin. The bottom floor was one large room viewing the lake on the backside and the road on the front. Stairs rising from the opposite side of the room led him up to two bedrooms in the loft, one larger than the other. Looking out the window of the larger one he could see the lake stretching like a dark mirror off into the moonlight. The other side of the lake was a dark, jagged line of trees blending into the ridge along which he had approached the cabin.

He came back to the bottom floor and turned his small flashlight into the cabinets. Within thirty minutes he had discovered a 243 Remington varmint rifle, a box of ammunition and enough insect repellent to last him a month. In the kitchen cupboards ample canned goods brought a smile to his face. He could stay here as long as necessary.

CHAPTER 21

Jason Amador brought the steaming cup of tea to his lips and studied Charles Burlingame through the hot mist. He had known Burlingame for a good part of their adult lives. He liked him as well as he liked any other human and respected him for his role as a father and community leader. He had been present when Burlingame was widowed and admired him when he had elected to raise the two children by himself. He had watched the stocky lawyer build a solid empire in the county and become its wealthiest and largest private landowner.

They were sitting in the lofty living room at L'Aerie, Jason's mountain retreat. Burlingame had called him with apologies and asked for an immediate meeting. He now sat before Jason, a grieving father—so lost in his own sorrow as to be somnolent. Charles moved with the stolid, drugged cadence of a sleepwalker. His eyes heavy-lidded, his speech slow and halting, bordering on the incoherent. Christa Bonamay, Jason's assistant, had handed Burlingame a cup of tea and left the room and he still held it as though unaware of its presence in his grasp. "Jason, I don't know where to turn—what to do. I've lost my whole family—the blink of an eye—I'm not sure I can handle it."

Amador put his tea down on the table next to his chair. He said nothing. He wanted Burlingame to continue talking, to snatch what solace he could from getting the hurt out of his system.

Burlingame drew a deep breath and let it out in shudders. Tears turned his eyes to red rimmed holes in his head. "The worst part is— the Sheriff, the GBI, the FBI,—none of them have a real clue to who did it or where the goddamned murderer is now. The Sheriff has had people running all over these hills since Kathleen—since it happened and The Woodchopper's still free—to do it again—and they all think he will. So do I..." His voice trailed off and he sat staring through the tea cup in his hand.

He looked up at Jason and for the first time seemed to focus on the present. "Can you please help me Jason?"

"Yes."

CHAPTER 22

L'Aerie is a habitat created solely out of Jason's genius. Living off private investments Amador is a millionaire many times over. He bought L'Aerie for practically nothing (considering its value to him). It was an abandoned grist mill cuddled against a mountain side overlooking a deep, clear pool in a free flowing sweet spring. The spring and pool were home to twenty inch rainbow trout. Three million dollars later it was state of the art in comfort, security, communications and travel with its own chopper pad on the hill just above the house and three satellite dishes artfully concealed in the attic. Tunneling like a mole back into the granite hill at the basement level he created a research scientist's dream lab, an ultra modern shop and a forty foot lap pool. To the remodeling price he added another like amount in a one-of-a-kind triple mode, voice operated, 15 Gigabytes Multi-Media Cray CPU (Central Processing Unit) plus every periphery available on the market and all built to his specifications. The giant two story water wheel whispering gently just below the bullet proof picture window guaranteed him a constant flow of power through two, 3-phase generators bolted to the floor in the corner of the workshop. Next to them a twenty ton heat pump draws its sustenance from the icy spring water before the icy water runs through the scrubbers and dissipates the damaging heat created by the super Cray as it calculates. The heat pump also keeps L'Aerie at whatever temperature is selected and keeps the lap pool at a constant eighty-five degrees. If the massive water wheel was ever disabled automatic diesel stand-by power in the basement romps into play. Couple that with the basement freezers bursting with food and a deep, sweet water well providing fresh water through the basement floor and the building is self sufficient for months if need be. Protected, as it is, with security measures of his own design not even a bird can approach within five hundred yards unannounced.

Supplies and the workmen to do the remodeling had all been flown in and out by helicopter so the locals were unaware of the scope and complexity of the premises. The only copy of the building plans was locked away in a three ton walk-in fire proof safe set into the hill side wall of the basement.

To drive by 'L'Aerie' on Highway 441 one simply sees a picturesque old grist mill surrounded by a high, steel, picket fence posted "No Trespassing" and fronted by a massive wrought iron gate supporting an eight foot eagle in full flight. The wing span of the eagle exceeds fourteen feet.

The electronics at "L'Aerie" responds only to two people. Himself—and Christa Bonamay.

Christa Bonamay, two inches taller than Jason Amador is a willowy, dusky brunette with large, lustrous blue eyes. Born out of wedlock in Jamaica to a stately Jamaican serving girl as a result of a wealthy Castillian vacationer's fling she spent the first ten years of her life watching her mother lay on her back for money and the next five years doing it herself. But cream always comes to the top and her brain wouldn't let her stay uneducated. She saved her pounds, her dollars and her marks and at sixteen bought a ticket to the United States, freedom and education. At nineteen she graduated high school with her head held high, and at twenty-two she walked away from the University of Miami Summa Cum Laude with the unlikely dual major of Psychology and Computer Science. She met Jason Amador while doing graduate work on full scholarship at Berkeley. He was forty-five, she twenty-four. He offered her a job, she declined. He accepted but told her to memorize his fax number.

Ten years later as a Senior Computer Consultant she decided it was time to add more spice and challenge to her life so she faxed him. He moved her to "L'Aerie" one week later. Her official duties to compile, edit and prepare for publishing the very private papers and essays of one Jason Amador. Her private, primary duty is to access and become familiar with the world's protected databases. She also manages to oversee the house and practice her inventive cookery on him, for both of which he is eternally grateful. Their personal relationship is a live study in symbiotic perspective—an emotional feeding trough from which they both sup. To him she is a brilliant protégé who accidently doubles as a housekeeper and daughter he will never have. To her he is a sensual mentor replacing the father she never knew and emotionally supplanting the lovers she wish she hadn't.

CHAPTER 23

Jason walked across the living room toward the ten foot sliding glass doors leading onto the deck. As he walked he said "Open doors." and they slid soundlessly back. As he cleared the doors he said "Close." and continued to the outer railing of the deck as the doors closed behind him. From his balcony the mountain dropped away below him into the dark blue of the shadowed pines. An occasional light blinking below him in the distance was the only indication he was not alone on the planet. Overhead the sky was a midnight blue canopy sprinkled with twinkling diamonds.

He leaned over the railing and studied the cold, dark pool thirty feet below. The only sound in the velvety stillness was the whisper of the perfectly balanced twenty-four foot water wheel revolving just below him. Then somewhere in the blackness a mourning dove aired its plaint and below him a rainbow broke the surface of the pool to take a fly.

In a conversational tone he said "Cray, give me Lights" and the pool, the upper stream and the compound around the house were bathed in light.

The pool immediately broke into deep ripples with an occasional fin drawing a straight line through the roiling water. He launched a handful of corn into the center of the maelstrom and reveled as one striper, more hungry than the rest, broke the surface to take a yellow grain before iridescing back into the dark water. He knew there was at least one matriarch down there over forty inches long who would probably make a new world's record if it was hung on someone's wall. It gave him a comfortable sense of power to know that they would stay in the pool as long as nature wanted them there. "Cray, lights off, please." he said.

It took a few moments for his eyes to adjust to the lack of light but it was compensated by a large, mottled moon peeping over the crest above the house. The barren areas on the hillside above him flowed gray black with silver edges punctuated by charcoal blotches stretching from the foot of each clump of trees. Mica in the rocks reflected a thousand tiny points like miniature fireflies frozen in place.

He said. "Cray" in a conversational tone.

"Yes, Mr. Amador." Cray answered in a pleasant female voice.

"Where is Peter?"

"Mr. Chaney is in San Francisco."

"Send him the following message with our usual encryption. 'Peter, meet the chopper in Atlanta. Give Cray your ETA soonest. End'"

"Thank you, Mr. A."

"Your welcome."

The moon hung for a moment at her midline than continued her slow climb into the dark sky. He watched the shadows shorten, coalescing, slowly changing. Somewhere out there is a killer, a killer who kills on contact. I'll set a thief to catch a thief. How appropriate.

* * * * *

"Christa, I want you to program Cray to sift all computer records, phone calls, faxes, and any other electronic references from the Police, Sheriff's Department, the Fish & Game Regional Office, local hospitals and doctor's records in all the counties contiguous to

Caravelle and Voodoo Mountain. Go back a week before the hikers died and make it ongoing 'till we get to the bottom of this."

"What are we looking for specifically?"

"Disappearances, odd sightings, unexplained injuries, break-ins, thefts, murders—anything out of the ordinary."

"Cray can do that by herself after I have a talk with her. What do you want me to do."

Amador smiled. He was always struck by the beautiful contrast between Christa's bright blue eyes and her tawny, milk chocolate skin. "I'm still amazed at you Christa. You had been here barely a week until you had programmed Cray to be a woman to me and a man to you."

Christa chuckled. "Fair is fair, Mr. A. What do you want me to do?"

"You get the fun part. Four things. First I want you to research Burlingame Enterprises, files and records. Look for disgruntled employees or ex-employees. See who stands to gain now that the children are out of the way."

He paused while Christa scribbled notes. "Second, I want you to crack the computer in the Corporation Department in the Secretary of State's Office in Dover, Delaware. I want to know who the principles are in "Ident, Inc." and anything you can find out about them. Get any names available of the personnel now in residence."

"Do you want me to follow the money, Mr. A."

"Definitely, I want to know where their money comes from…"

A dulcet voice interrupted him. "Mr. Amador."

"Yes, Cray."

"Mr. Chaney just gave you an ETA of ten a.m. tomorrow morning here."

"Thank you, Cray."

"You're welcome, Mr. Amador."

"Is that voice getting sexier each time I hear it or is it my imagination?" Jason asked. Christa could hardly contain her laughter. In fact she put her hand up in front of her mouth to stifle it.

"She's a work in progress." She said.

"Well I'm glad we now have something of greater importance to occupy your vast talents…"

Christa giggled then returned to business. "You still have to give me number three."

"Yes—I want you to find out which Atlanta Surveyor did the Voodoo Mountain work. Get into his files and download the final plats of Voodoo Mountain. Number four, download the latest geodetic satellite survey of this area. Get an infrared one also if its available." He stopped and snapped his fingers. "Christa, there must be satellite photos showing their choppers at work when they developed Voodoo Mountain."

She nodded, making notes. Jason continued, "Go back to just after the mountain was sold. Tell Cray to give us copies of any activity over the Caldera from that time to this. We know the inside of that mountain is pocked with caverns but they had to dig down into them and build the buildings on top. That takes large, powerful equipment."

Amador paused for a moment, lost in thought. Christa had learned early in their association to wait out these pauses so she kept silent.

"A fifth item just occurred to me. See if you can find the architect that designed that underground facility up at Voodoo Mountain. If you can, download a copy of their plans. Look for any part of the construction you can find—maybe suppliers. I know all that material didn't come from overseas."

He walked over to another console and sat down. "Let me know as soon as any parts of those come in. Now if Cray will still talk to me I will set up a tri-level scanner to read any messages coming out of Voodoo Mountain."

"Of course I will, Mr. A." Cray answered with a lilt in her voice.

"All right Cray, here's what I want you to do. I want you to plot Voodoo Mountain on the GPS (Global Positioning System) and implement a tri-level scan. Record any message in or out. You'll have to translate most of them into English but you've got more than 3000 languages in there to choose from so it shouldn't be a chore for you. Right?"

"Of course not, Mr. A."

"Good. They'll also probably be in code so I want you to decrypt them, then give us a hard copy—and I want you to fly on this one. ok?"

"I have only one speed, Mr. A." Cray quipped. Jason could swear he detected drollery in Cray's tone but dismissed it as impossible.

"Yes, and I want every last Hertz of it." He said determined to have the last word.

"That's what you pay me for, Mr. A." was Cray's retort. Jason heard a muffled guffaw and looked sideways at Christa who was studiously bent over her console, her shoulders vibrating with restrained laughter.

CHAPTER 24

"Christa?"

"Yes, Cray."

"Mr. Chaney is requesting permission to land a helicopter on the pad in ten minutes."

Christa's heart rate increased a couple of beats. It had been some time since she had seen the tall, dark and handsome Chaney. She liked him enough that this visit would add some extra spice to her life. She knew Peter well enough to know that he would be in rut all the time he was near her at L'Aerie. She smiled and thought for a moment. "Cray, focus the main sensor on the pad, put it on the big screen and notify Mr. A."

"Done."

It still amazed Christa to realize that this monster array of chips and wire could interpret voice commands so readily and perform them immediately. In her career she had worked with some very advanced and even quantum-step experimental computers but nothing had prepared her for the sophistication of Cray.

As Cray finished speaking the outside view of the chopper pad and the surrounding hillside burst into living color on the huge six by eight foot wall monitor.

Christa was in the heart of L'Aerie, the computer room, jokingly dubbed the CPU by Jason, which was actually a forty by sixty foot basement blasted out of the solid rock of the hill. Its four walls were rock and the floor and ceiling were poured cement. Inside was cool and quiet as a tomb, the only sound being the soft sighing of Cray and her accessories going about the appointed tasks. At one end were the doors leading to the lab and the workshop, the walk-in safe and the lap pool At the other end leading to the engine room a ten foot double steel door hermetically sealed as it closed, blocking any fumes or temperature changes from transmitting from the engine room to the CPU.

Jason came through the door from the workshop.

"Cray."

"Yes, Mr. A."

"Let's see how good you are. That chopper has to be at least five nautical miles SW of us. See if you can pick her up on the big scope."

"Elementary, Mr. A. I have been locked on the signal since Mr. Chaney's transmission."

Jason chuckled. "Well then, Cray you seem to have everything under control. Perhaps you can tell me who manufactured that chopper and what model it is."

The scene on the huge monitor shifted to reveal the valley flowing away below them. A red circle appeared in the sky on the upper corner of the screen and slowly moved down toward the center.

Cray said. "A McDonnell Douglas MD 500 E single turbine four seater, Mr. A. Would you like its specifications?"

"No thank you Cray. Please zoom in slowly."

"Yes Mr. A."

The circle grew steadily larger until a chopper became visible in its center. "Cray, print the range in yards on the screen and update it every twenty seconds."

"Yes, Mr. A."

The figure 7004 popped up on the lower edge of the screen. Christa sucked in her breath. "Good Lord, nearly four miles out."

Amador smiled. "Her maximum working distance is about ten miles give or take a few yards. Watch this Christa., Cray."

"Yes, Mr. A."

"Go to maximum resolution and focus on the passenger."

Jason turned to watch the monitor as the passenger's face came into full screen, then stepped forward in disbelief. "That's not Peter, Cray, where's Peter?"

"Mr. Chaney is in the pilot's seat flying the helicopter."

Jason smiled. "Getting in some air time, eh. Cray, focus on the pilot."

The screen blipped and the pilot's seat came on screen. Through the windshield Peter Chaney's darkly handsome face came into focus. Christa's heartbeat quickened.

Jason flicked a switch on the console. "Peter, can you hear me?"

Peter's face broke into a huge grin. "You bet, Mr. A. Loud and clear."

"Welcome back to L'Aerie. Did they get my supplies on that chopper with you?"

Peter jerked his thumb at the rear of the cabin. "Thanks, yes. We've got four or five boxes on board. We're loaded to capacity but I don't know what's in 'em."

"Time enough for that when you get here."

"Right. Is the lovely Christa there?"

Jason smiled and raised an eyebrow at Christa. She said, "Yes, I'm here. It will be nice to see you again Peter."

"Likewise. We're about half a mile down valley now so I'm going to give the stick back to the pilot. See you both on the ground."

CHAPTER 25

"Peter, off load the boxes to the pad and let the pilot go. I'll see you up there."

"Right."

Jason flicked the mike switch to off and spoke to the computer. "Cray, put the main sensor on the pad and run the scope back to normal."

"Yes, Mr. A." As Cray spoke the big screen displayed the chopper pad and the surrounding area. The chopper was just coming into view in the distance. The pilot brought it to hover just off the hillside, then slid sideways over the pad and lowered the bird with great care. It came down slowly, gently waddling side to side like a giant mallard hen settling onto her clutch of eggs.

Peter stepped out from the left side and began taking boxes from the chopper and stacking them on the upper edge of the pad. Christa watched the tall, lithe figure slightly hunched over move easily back and forth under the whirling blades and wondered what it would be like to break her self imposed celibacy and surrender herself to this man. From her first meeting with Peter she was physically drawn to him but had maintained a position of 'armed truce' as she did with all men. She let her mind wander for a moment back to Jamaica and she cringed inside. The harsh truth was that on her island sex was the renewable survival commodity. If you were a beautiful, ambitious young woman there was only one way to move from trash to class. So for many years as a girl-woman she steeled herself to trading sex for money—always hating it—always tainting the act with the shame and—not being able to bear the shame—in penance had abstained from sex since leaving her island. It was obvious Peter desired her as did most, except for Jason. Jason did not appear to desire any woman nor need sex of any kind.

She glanced sideways at Amador as the two watched Peter on-screen. Jason, oblivious to her scrutiny, polished his glasses. You have become the consuming passion in my life, she thought, my aging genius. A pity—you are as naive and unaware of that fact as a three year old is his nudity. She turned back to the screen. Jason broke the silence. "Christa, how about digging up some sandwiches, we're all going to be hungry." Christa nodded as she watched the screen.

Peter gave the pilot 'thumbs up' and shielded his face from the wind. The chopper lifted a few feet into the air, backed away from the pad and in moments drifted off-screen.

* * * * *

During remodeling of "L'Aerie" Jason installed a four story lift through the solid rock from the floor of the workshop to the helicopter pad. He cut the pad level into the hillside which left a cliff face of solid rock on the upper side of the pad. In the cliff face were three doors cut in the rock. One a storage room for outside gear, another to a set of stairs leading down to the workshop and the third the door to the elevator down to the workshop. The elevator and the stairs can be entered from the living quarters at any one of the three floors.

Jason got a hand truck in the shop and took the lift to the chopper pad.

The two men shook hands and Peter threw his arms around his mentor. Jason, obviously ill at ease, endured Peter's spontaneous show of affection, but finally shrugged off the larger man, stepped back and readjusted his glasses. "Well Peter, I see you're keeping fit."

For the moment Peter was a large puppy gamboling about the pack leader. "Never know when I'll be needed Mr. A. It's been six months since I've been here and I've missed you."

Jason smiled. "Well we've certainly got something to your liking here now, Peter. Let's get these boxes below while I tell you about it. Then you can go into the kitchen and help Christa with the food."

* * * * * *

Peter grinned in anticipation and started throwing boxes on the dolly.

Before he rounded the final corner into the immaculate kitchen Peter could hear the clink of utensils, the thunk of the refrigerator door closing and the myriad sounds a woman makes holding sway in the kitchen. He peeked around the corner.

Christa was standing right side to him at the counter, lathering mayo onto slices of bread. The tall, dusky brunette was wearing tight, black toreador pants and a short printed blouse tied just above her bare navel. Her high tipped bosom swelled into the blouse as she breathed and Peter stood entranced, as he watched, barely able to draw breath. A tiny drop of moisture clung to her upper lip like a bead of dew in the morning sunlight. She licked it off. He stepped into view.

"Oh," she said, flicking an imaginary hair out of her eyes, "you startled me."

He smiled smugly, "Jason told me you were down here. I thought I could help you do the sandwiches."

"Somehow I don't see you as very effective around the kitchen."

"Who do you think feeds me when I'm out in California?"

She laughed. "Oh, I don't want to open *that* can of worms."

"Really," Peter rallied. "I can cook. I'll admit I don't do it very often but I can when necessary."

She rinsed the knife off. "What have you been doing with yourself out there?"

Peter was diffident, ill at ease. He wanted to score high with this lady but hadn't the slightest idea how to do it. She gave no clues and he knew from experience she would feed him to the sharks in a moment. He picked up the sandwich plate and inspected the sandwich with inordinate care. "I bought a vineyard."

She turned and gave him her full attention. "You bought a bunch of grapes?"

"No, I bought a piece of a vineyard. I bought the vines from which the grapes come."

She laughed. "Real live grapes. Real live vines. What do you know about grapes?"

Peter grimaced. "You mean what does an ex-con, practicing psychopath know about grapes?

It was her turn to fidget. "No, I, really I didn't mean…"

Peter grinned and relaxed a little. "In the beginning it was just a lark, a way to make a little tax money back. But I've been getting into it and now that I know something about them I'm enjoying it."

He looked at her. She was interested. Her blue eyes were so intense. He said, "You know the part I like best?"

"Tell me."

"Hit the vineyard early in the morning, at first light. The dew is shining on the leaves and it glistens on the grapes. There's no other sound but in the background you can hear a whippoorwill or maybe a bobwhite…"

"…We hear them here."

"…It's very—satisfying." he added lamely.

She smiled, a little cat smile. She asked. "What kind of wine?"

"Chardonnay."

"I love Chardonnay. Does Jason know about this?"

"I haven't mentioned it to him yet."

Her laugh tinkled. "You should when this is all over. He's a walking encyclopedia of wine lore. I don't think he's ever touched the stuff—the most he drinks is beer—but he can tell you all about it. You must send us some, maybe I can get him to try it…"

Peter took her hands and shook her arms gently. "Would you hush for a minute."

She looked up innocently. "Why Peter, you know that when a lady is ill at ease she talks overtime to cover it up."

"Are you ill at ease?"

"Well yes, you've got that serious look in your eye, and that makes me uncomfortable. I don't think you want to do that."

Peter held her close and looked down into those startling blue eyes. They were deep blue pools—you could lose yourself in there.

"You know that I could really care for you if you just gave me half a chance.

She drew away from him and looked long into his brown eyes. "Peter, dear, I hear what you say and I feel what our bodies say when you get close to me,—I want to respond, but my mind says I'm not ready yet." She moved to the end of the counter and pulled three beers from the refrigerator. He watched her lithe movements hungrily. She turned back toward him her eyes glistening with unshed tears. "Those five years I spent as a hooker were hell on earth. It was my only way out but I hated every moment and I hated every man who screwed me…" She clenched the bottles of beer until her knuckles went white and Peter thought the bottles would shatter. Large clear tears rolled down her cheeks as he stepped forward and gently took the bottles from her. Embarrassed, she turned away from him muttering, "…and I still can't wash the stench off me." and disappeared through the far door.

Well, that will teach me to open my big mouth. He picked up the food and went back to the CPU.

* * * * *

Jason lifted one of the sandwiches and took a greedy bite. With his mouth full he asked, "Where's Christa?"

Peter munched for a moment,—he was not pleased to discuss it. "I upset her," he answered reluctantly.

Jason took a long pull on his beer and reflected on this. "That's not easy to do."

"I accidentally brought back some bad memories."

"Bad memories leave bad scars." Jason said gently and reflected. "Those scars reshape us and handicap us emotionally. In order to live a 'normal' life we have to learn to live with the new constraints and ignore the disfigurement. To look at that gorgeous brown belle you'd never figure her for much baggage but once in a while the stench comes to the surface and she hasn't totally come to terms with it yet."

"Funny," said Peter. "That's the word she used—stench." As an afterthought he asked. "Do you think she ever will?"

Jason was preoccupied, watching one of the screens. "Will what?"

"Do you think she will ever come to terms with it?"

"Yes, I expect she will. However, it's going to take a world of gentleness, kindness and understanding from both of us to do it."

"I haven't had much practice in those fields, Mr. A."

Jason smiled. "Patience, Peter. The act of living almost always allows us to acquire the needs for survival if we will but pay attention and learn."

"It certainly happened that way in my life." Peter agreed.

"Yes, and if you continue to absorb what life has to offer you may someday acquire those other qualities that will unlock the lady's emotions and bring her willingly into your arms."

"You sound more optimistic than I feel."

"I have great faith in you Peter."

CHAPTER 26

"Colonel."

Colonel Pong-Yul looked up from his desk. "Yes, Lieutenant Pong-Uk, what is it?"

"Sir, I am getting some strange interference in the wavelength between us and the satellite."

"What does that indicate to you?"

"I believe someone is trying to monitor our transmissions sir."

"Can you trace it?"

"It is on more than one level, Sir and it switches levels erratically. This leaves me practically no pattern to…"

Colonel Pong-Yul broke in, his words trailing an icy edge. "Can you trace it?"

The lieutenant chose his words very carefully. "Only with time, Colonel Sir."

"How long?"

"Perhaps a week Sir, maybe ten days, Sir if they keep transmitting."

"Lieutenant, I do not have ten days. Our work may be finished here in less than seven days and to accomplish that I will brook no interruptions,—none. Do you understand?"

"Yes, Colonel Sir."

"Until you trace those signals go to level two encryption."

"Yes, Sir."

* * * * * *

"Dr. Ho-Sik, this is Colonel Pong-Yul. Can you spare me a few moments?"

Ho-Sik grimaced at the phone but his voice was calm and level. "Yes, Colonel. Do you wish to talk to me over the phone as we are doing or do you want to come down here?"

Pong-Yul replied. "Just a few questions, the phone will suffice."

Ho-Sik's sigh of relief was real but inaudible. He did not really want to waste time with this half breed pouter pigeon. "What do you want to know?"

"The last time you and I talked you said our subjects were all ready physically but not sufficiently schooled yet. Where do we stand now in being fully ready?"

Ho-Sik looked at a series of charts on the wall opposite his desk. "Eleven are ready now. The other two need a few more days indoctrination."

"This is Wednesday, Dr. I have communication from Our Supreme Commander in which he states unequivocally he wants your subjects ready no later than Monday."

"Colonel these things take time and I'm not sure we will be able to meet that...."

Pong-Yul cut him off. "Dr. let me be candid. Have you been in the lab so long that you have forgotten who your masters are? When the Supreme Commander says he wants us ready Monday it is worth our respective lives if we are not, do you understand?"

"They have no idea of the complexity of this..."

The Colonel exploded. "Complexities be damned, Dr. I have a long and distinguished career with the NKPA and I for one am not going to literally lose my head over this project. I have no choice—you have no choice, you must have them ready to fly out of here on Monday."

Ho-Sik caught the breath of hysteria in Pong-Yul's outburst and was experienced enough not to confront him head-on. He knew this was one of those encounters in which one side teeters on the brink and any confrontation will send the incident into orbit. He backed water. "I think I can guarantee you they will be ready when Our Supreme Commander wishes to deploy them, Colonel. Now I really must get back to work."

"Yes," Calmer now, "but keep me posted if there is to be the slightest deviation from the schedule."

"You have my word, Colonel."

CHAPTER 27

Christa downloaded the latest black and white and infrared satellite geodetic maps of the area and then blew up the small sections until she had a map four by five feet that covered an area roughly fifteen miles north to south and twenty miles east to west.

She spread the map on the only table in the shop that wasn't covered with parts for the motion/light sensors which Jason and Peter were assembling. Jason discussed a couple of small points with her and ignored her gruff answers. She did not look at Peter but went quickly back to the CPU to commune with CRAY.

Peter and Jason went back to the sensors. Brown in color, sensors, an Amador invention, were photo voltaic devices sensitive to both light and motion and were adjusted to register the movement of anything larger than a raccoon. Each was one and seven eighths inches in diameter and looked like a mahogany radiator cap with a one inch hardened steel spike on the backside.

Jason stepped over to the map and tapped it with his fore finger. "Look Peter. The WoodChopper murdered the two hikers somewhere

121

up here and dumped the bodies in the crevasse. Then he was sighted over the ridge, here, in Cherokee County when he stole the groceries. Next, he murders the two clerks down here at the convenience store. Now where do you think we aught to look for him?"

Peter grinned, "That's too easy, Mr. A. We look for him where he hasn't been."

"Of course, now that he has a map he can go any place he wants and so I think he'll head for this area up here around the Devil's Play Pen. Besides, there are houses up there and he can steal food."

"Some of those houses are unoccupied, aren't they?" mused Peter.

"Yes." Jason pointed to the foot of Trail 1. "Here's what I want you to do. Jog up Trail 1 to the head, cross over to 2 and come back to 441. Then take the lake road to Trail 4 and go back north. About every hundred yards I want you to spike one of these sensors into the southeast side of a prominent tree as high up as you can comfortably reach."

Peter turned one of the sensors over in his hand. "I've been building these things for an hour now, tell me more about them."

"Each has a chip, a crystal and a battery inside. The circuit is connected to the spike and when you 'plant' the spike in the tree the chip is activated."

Peter raised an eyebrow. "What do I 'plant' them with?"

"Actually you just slap them into the tree as you pass by." Jason smiled. "By the time you've done three or four of them you'll probably be doing it at a dead run. Anyway, Each has its own distinct signal which pulses nearly half a million times a second. That pulse covers a field of one hundred seventy-five degrees out about a hundred yards in front of the sensor. When something moves through that pulse that image beams to the satellite and back to Cray here. Cray interprets the signal and gives us a visual on the monitor. That signal also tells us which sensor we are dealing with." As an afterthought he said. "I gave them the brown color so that no one would notice them and so we don't have to go back and take them out."

Knowing he would be the one to purge the woods of the sensors if it was necessary Peter drew a sigh of relief. He started putting time and motion together in his mind. "I'm in good shape, Mr. A, but this is going to take the better part of two days."

Jason walked over to a wall cabinet and came back with a strange looking pair of thick-soled jogging boots. He handed them to Peter. "Not with my Striders, you still wear a twelve?"

"Sure, what do we have here?"

"This is the latest pro basketball shoe available. Take a look at the sole."

Peter put one boot down and concentrated on the other. The sole was about an inch and a half thick and spongy to his touch. The surface of the material was tacky and indicated excellent traction on hard surfaces like rocks.

"Put it down on the table, drop your full weight on the top of the shoe and see what happens."

Peter put the heel of both hands on the shoe and dropped his full weight on top of it. The shoe mashed flat under his two hundred and twenty pounds and the sole condensed to half its height and then rebounded. The force threw him into the air and backward. He doubled up his knees instinctively and landed like a cat five feet from the table still clutching the boot.

"Good Lord!"

Jason was pleased with himself. "With a little practice you'll be able to control your movements and should be able to cover up to three meters at a stride in open country."

"What gives them the bounce?"

Jason took the boot and traced his finger along the sole. "I redesigned the sole and inserted an adjustable polymer spring between the two halves. It's calibrated for your weight now but can just as easily be set for mine or Christa's."

"What about weaponry?"

"I've been having some fun experimenting," said Jason. He opened a drawer and brought two weapons up to the table. He picked up one which looked like a standard rifle stock about thirty inches long with a butt plate on one end and a metal slot that ran the length of the stock on one side. He rotated the butt plate, tipped the stock down and out slid a thin barrel as big around as your thumb and a distance compensating, laser scope. He inserted the barrel into the end of the slot and rotated it ninety degrees to lock. He slid the scope onto the grooves at the side of the chamber and locked it into place. The whole assembly had consumed thirty seconds. He gave it to Peter. Peter whistled softly when he looked at the bore. He could almost stick his thumb into the hole.

"The barrel is standard heavy duty rigid electrical conduit." Jason said. "The whole weapon only weighs three and a half pounds." He handed Peter a long oddly shaped cartridge that looked like a storage tube for a small cigar. "This weapon is based on the old Gyrojet technology from back in the sixties. It is really a small diameter rocket. You simply insert the rocket in the rear of the tube, push it forward until it clicks, then aim and push that little button there." He indicated a small black button on the left side of the stock within reach of the shooter's right thumb. "With these special hand loads of mine you can hit what ever you can see at five hundred yards. There's only two negatives."

Peter took his eye from the scope and looked at Amador. "Oh?"

Jason handed him four more cartridges. "Since it is a rocket there is blow back and your target can trace the shot. Velocity is around twenty-two hundred feet per second so they shouldn't have too much time to react. But be sure to keep your parts at the side of it, not behind it. If you have to ditch the weapon, push the little button here on the butt and in thirty seconds a small thermite unit will consume it—and about everything else combustible around it."

"No plinker, eh?"

"No, now this one is a sweetheart." Jason said, hefting a futuristic long nosed pistol, probably eighteen inches long from ugly snout to end of breech. "I started with an Anschutz frame and modified it. Now practically everything but the firing pin, breech and the barrel insert is polymer so it only weighs thirty ounces without the scope. The barrel is eight inches long and the rest is compensator and silencer. The scope is a red dot, dead on at anything up to two hundred feet."

"It's thirty caliber isn't it. How many rounds?" Peter interjected.

"Yes, shoots standard 30 caliber carbine ammo and the magazines hold thirty rounds. Single shot or three round burps. There's the lever," he indicated a small flip lever near the right thumb depression.

"However you will be going out with something special in the way of ammo."

He reached back into the drawer and brought out two boxes. He selected two cartridges, one from each box and held them up to the light in his left hand. "This one," and he pointed to the left one, "is the standard fully jacketed 30 caliber carbine. One hundred ten grains, with a velocity of about nineteen hundred and eighty feet per second. But this" and he took the right one and held it up to Peter. "This, one is totally my own design."

Peter could see the cartridge was slightly shorter but the bullet was longer than the other. He hefted it in his hand. "Heavier, isn't it."

"Yes, one hundred eighty grains to be exact. Slightly less velocity, the big difference is in the explosive charge."

Peter pursed his lips.

"Yes," said Amador. "This bullet explodes after penetration. This cartridge will behead your target—but it makes a loud noise as it does it."

"How tender is that ammo, Mr. A?"

"It's armed by firing, until then you can handle it like ordinary ball."

Peter pushed the rounds back into their boxes and queried. "When do I go into the woods?"

"Saturday morning before first light. That just gives you tonight and tomorrow to familiarize yourself with this and the other equipment and get used to the boots. Now let's get these sensors finished."

CHAPTER 28

It was decadent luxury sleeping on a mattress after so many nights dozing fitfully in the mosquito ridden woods.

The waves of pain were more frequent now, the nausea more pronounced, the lethargy deeper. Each time his body went under siege a little part of him slipped away and did not come back. He brought the mattress down from the main bedroom into the living room and dropped it on the floor in front of the picture window so that he could monitor the front of the cabin facing the road. It was well he did. Flashing lights woke him at four a.m. on Friday morning.

The Sheriff's Patrol Car came slowly up the country lane probing the woods and surrounding areas with a halogen spotlight. It pulled into the yard and stopped. Two large uniformed figures stepped warily from the vehicle, drew their weapons and walked carefully toward the cabin. They split, with one staying in the front, the other working his way to the back.

He folded the mattress in his arms and pushed it behind the breakfast bar out of sight and hunkered down beside it. He heard one of the deputies mount the steps to the front deck and ghostly shadows flitted around the furniture as a powerful flashlight reached into the four corners of the room. The other deputy tried the lock on the back door and The WoodChopper held his breath. When he had entered the cabin he had gently jimmied the door with his survival knife but had left it locked. It held. The deputy walked the length of the deck rounded the corner of the cabin and poured his light into the room from the lake side. Both flashlights turned the cabin's main room into daylight. He held his breath. It seemed like forever that the two bright swaths of light held, swayed from side to side, then held again. Finally the lakeside beam pulled from the room and he watched it recede toward the lake. The beam on the front moved out of the window, floated down the front steps and in a few moments was joined by the other and the patrol car backed into the lane and drove away.

He let his breath out in short gagging bursts. Sweat was flowing from his forehead, dripping into his eyes, coursing down his cheeks. He wiped his face angrily, stunned by how close he had come to discovery, It was obvious that if he stayed here he would be found.

He gathered as much food and water as he could carry, strapped the rifle on his back and melted into the north west woods.

* * * * *

"Mr. A.?" Came from a wall speaker.

"Yes Christa."

"Some reports you need to see in the CPU."

"O.K." He turned to Peter. "I'll be back as soon as I can, stay with this and the rest of your preparation." Peter nodded as Amador left the room.

She handed him a sheaf of papers. "Cray broke the original Voodoo code quite easily, except for destination data. It was straight out of the North Korean Code Book but now they've changed it..."

"Which means they're on to us."

"Yes, it would appear so." She indicated some color on the sheets. "I've highlighted the wheat, and there's very little, the rest is chaff. The local Commander's name is Pong-Yul."

Jason scanned the sheets. "Right, nothing here to help us. Cray!"

"Yes, Mr. A."

"What length of time are we alternating the scanner beams now.?"

"Twenty seconds."

"That is apparently too long. If you were to randomly select times between five and fifteen seconds will that give you enough time to maintain full contact with Voodoo Mountain."

"Yes, Mr. A."

"Do it please and monitor their probes and keep me informed of their progress in locating us."

"Yes, Mr. A."

He turned back to Christa. She said. "Some significant news from the local authorities. First—the autopsy reports on three of the dead. Kathleen and the two hikers. The sheriff did not tell her father that the killer urinated on her after dragging her into the storage room and on the hikers before throwing them into the crevasse."

"Extreme symbolism. So now we know there is no doubt that The WoodChopper is the killer of all. What we don't know is why he does it—and we may never know. What else?"

Christa shuffled the papers, found the one she wanted and said. "As you know Calla County is directly north of Voodoo. Two hikers, man and wife went in for a one week camping trip. Should have come out three days ago. No word, now presumed missing."

Jason looked at the big wall map. "How far above Voodoo Mountain did they go in?"

"Forty miles."

Jason looked at Christa. Both nodded. Jason said, "that's too far away to be The Woodchopper, he couldn't be every place at once. I do not believe in that kind of coincidence, more than likely there's someone else out there. If they are from Voodoo Mountain they are looking for him just like we are—and probably desperately want to get him before we do."

While he was talking Christa was rapidly shuffling through the papers in her hand. She pulled one out and thrust it at Jason. "Before they changed the code Cray picked this up. It's the Officer of the Day's Log. They must have to transmit it *in toto* to wherever headquarters is."

Jason scanned the log. "They sent out four squads on recon. Wonder how many they usually have out?"

Christa thought for a moment. "I remember them saying some place that 'the extra squads were deployed in light of current requirements'…"

"What else?"

"Burlingame Industries has no contingency for loss of heirs. At this point, nobody but the state benefits. And I find no records of angry or disgruntled employees. Au Contraire, every body seems to be happy with Burlingame."

"Christa."

"Yes, Cray."

"I have compromised the new code from Voodoo Mountain."

"Marvelous, continue to monitor them on the same parameters as previous."

"Miss Christa."

"Yes, Cray."

"I have just downloaded the requested Corporate files from the Secretary of State's Office in Dover, Delaware and have broken contact."

"Cray, is that the information on "Ident, Inc" that I requested?"

"Affirmative."

Jason and Christa looked at each other. Christa shrugged, "I agree, that's a waste of time. Cray, in the future will you please substitute 'yes' for 'affirmative' and 'no' for 'negative'.

"Yes."

Jason broke in. "Cray."

"Yes, Mr. A."

"Christa just gave you a command on substitution of words, right?

"Yes, Mr. A."

"Those substitutions will hold true except when you are discussing film. When you are discussing film, a negative is still a negative not a 'no'. Can you make that distinction, Cray?"

"Yes, Mr. A."

"Thank you." He said, making a face at Christa.

She said. "Cray?"

"Yes?"

"How many pages in the Delaware download?"

"Sixteen."

"How long were you in their files."

"Elapsed time was four hundred twenty five thousand two hundred fifty seven nano-seconds."

"Well under half a second," Jason said dryly.

Christa smiled. "Cray, You closed their files and left them as they were?"

"Yes."

"Cray, did you detect any reaction from them as a result of your intrusion."

"No."

"Thank you. Cray, put thumbnails of all sixteen pages on screen at one time and print me a hard copy of each."

A printer began a quiet purr in the corner and the big wall screen came alive with pages of text and contracts. They began to scan the pages.

"Here's the owner," said Jason pointing to one page. "His name is Khan Ree and he's listed as a Korean National. His representatives here in the United States are John Wayne and Gabby Hayes. There's a dead end if I ever saw one."

Christa Chuckled. "At least they have some imagination."

"Well, that's why Delaware is used to incorporate so you can do it legally without scrutiny."

"Miss Christa."

"Yes Cray."

"Message volume on Voodoo Mountain has increased three hundred percent in the past fifteen minutes."

"Cray, display those transmissions in thumbnail please."

The big screen blinked and the Delaware paperwork was replaced by messages from Voodoo Mountain. Jason and Christa began scanning them.

"We've cracked the destination codes," Christa said breathlessly, pointing at the screen. "Look, they are all going to Tokyo to a company called Ting Sao."

"Cray,"

"Yes, Miss Christa."

"Search Ting Sao, begin in Tokyo and go world wide."

Jason was reading the messages. His face was grave. "There is something very serious going on down there, Christa. Colonel Pong-Yul has only just told Headquarters that G-101, that's The

WoodChopper's code name, is missing. All hell is breaking loose at Headquarters. They're saying that all must be in readiness by Monday—but they don't say what…"

"Miss Christa."

"Yes, Cray."

"Ting Sao is a Singapore based off shore investment company with large interests in Japan. The message destination in Tokyo is in a building owned by them but is merely an electronic relay station."

"Cray, where does that signal go from Tokyo?"

"To satellite."

"Cray, track it from there."

"There are millions of messages transmitting through that satellite Miss Christa. It is impossible to pick out just one."

Jason broke in. "Cray."

"Yes, Mr. A."

"Intercept the messages from Voodoo Mountain and attach the same short nonsense signature to each before you put it back on line. Then look for and track that signature to its destination and report back to Christa."

"Yes, Mr. A."

"Brilliant," said Christa.

Jason allowed himself a smile. "That's what they pay me the big bucks for."

* * * * * *

Jason opened the door to the workroom to see Peter, in full battle regalia bounding down the length of the room. The workroom was a full sixty feet back into the hillside but with a ceiling of only ten feet it was taking all of Peter's marvelous agility to bound into the air and try for distance with out scraping his head on the ceiling. He stopped when Jason walked in.

"How do you like that camo, Peter?"

"First class, Mr. A. I should just disappear in the woods"

Peter was wearing a 3D camo suit which consisted of one inch squares of camouflage material shaped like leaves sewed one edge only to a green nylon mesh background. That nylon mesh background was stitched onto a Kevlar vest. With headpiece, body and gloves matched to the surrounding woods colors, he would disappear once he stepped into the woods.

"How good is the vest?"

"It will stop up to a 357+P^5 round. "Course, if you're not braced it will knock you down and probably take the wind out of you. Have you tried the infrared scanner yet?"

"No, I know it's working but I'll have to get outside for a true test. Talk to me about this big wrist chronograph."

Jason picked up the monster watch, about three inches in diameter and cleaned the face against his shirt. "This does a number of things, Peter. It emits a silent pulse once a minute that feeds to the GPS and into Cray so we know exactly where you are on the map at all times. The crystal face is also an LCD^6 screen so Cray can send you real-time readouts from your two infrared scanners…"

"Two scanners?"

"Yes. Two scanners and a miniature radio that connects you to Cray." Jason picked up a short one inch wide nylon strap with two widely spaced scanner units attached which looked like thick black coat buttons. Between the scanners an ear plug was suspended from the strap and from the ear plug came a transparent voice tube. "This velcros around your head under that hood. These scanners give you low level three hundred sixty degree coverage up to a hundred yards. The ear plug goes in your ear and the voice tube fits off the corner of your mouth. When you want to talk to Cray you simply say 'Cray' and then start talking. Everything said over the system is encrypted on input and decrypted on output by Cray so our conversations are secure."

He put it down and picked up another unit much larger shaped like a small walkie talkie and having two velcro straps dangling. Peter's eyes followed him like a hawks from within the headpiece. "This one attaches to your right wrist and you can zero in on a man sized target up to half a mile line of sight. Both read into Cray and Cray interprets for you. Cray will give you verbal readouts through this little ear piece or visuals on the face of your watch, which ever you prefer."

Peter whistled. "Looks like you've thought of about everything Mr. A."

"That's my part of the job, Peter, and it's the easy part," His voice turned grave. "but you've got some hard sledding ahead of you."

Peter's face split in a lop sided grin. "I've always liked the unknown, Mr. A.—this is right up my alley. But I need mission instructions. Give me an idea what I'm dealing with out there and what you want me to do with what I find."

Jason hoisted his rump up onto the table, leaned forward with his arms spread wide on each side of him. "You already know about The WoodChopper. I'd like to have him back here, preferably alive. We don't know his nationality but it is probably North Korean National. I do know that the NKPA is out there and you can treat them as the enemy,—but be aware that they will view you the same way."

A wisp of a smile crossed Peter's face but his eyes were cold pits of ice within the headgear.

Jason folded his arms and continued. "We're riding two horses here Peter. I started into this project to help a father assuage his grief, that's all. A man loses two children to a brutal butcher he has a right to look that bastard in the eye. That's the only way he'll ever have a chance to put the grief behind him and get on with his life. But the farther we get into this thing the more it looks like we're getting mixed into something that may be as big as our own national security."

Peter stripped the headpiece off his head. His hair was mussed and standing on end. "You be concerned about those things, Mr. A. I'm just going out there and have me some fun. What other back up do I have out there?"

"Whatever other weapons you want to carry out of the arsenal, Peter. Of Course I've got Condors[7] ready to go and Christa and I will be constantly monitoring." He picked up a tube about the diameter of his thumb and twice as long. "Christa has found you these high energy tablets and I want you to take enough MREs[8] and water to last you for a week, although I don't see you out there past Monday."

"Is that my deadline?"

Jason slid off the table and walked to the wall map. "No, but something is happening down there," he mused, tapping the middle of the caldera, "and it all points to Monday as being the big day." He turned to Peter and put his hand on the bigger man's shoulder. His quick, blue eyes bored into Peter's dark ones. "I want you to be careful Peter, and I want you to come back to me alive. We don't know their strengths nor weaknesses. We can only guess at how many there are. Considering their messages they have to be NKPA and that places them here illegally. We know they're up to no good but we don't know what it is. For this kind of covert mission in enemy territory they would use an elite cadre filled with totally dedicated men. These soldiers will ruthlessly butcher anyone who gets in their way so if you run afoul of them you'll get no quarter. Remember the four kids"

Peter's face went hard and his normally quizzical dark eyes turned to glistening chips of black diamond. In his briefing, Jason had shown him police photos of Kathleen Burlingame and Lester Westmore taken the night of the convenience store murders. He would carry those gory scenes in his mind till his dying day. Jason shuddered as Peter's smile morphed into the malevolent rictus of an evil gargoyle and a murderous haze clouded his eyes. The spasm passed in one long moment, just a flitting movement on his face but long enough for Jason to realize that the monster within Peter still lurked just beneath the surface. Peter smiled, a lop-sided, even toothed grin. "I remember them, Mr. A. Perhaps I can even the score—but can't you ring the local police into this?"

Jason shrugged his shoulders. "What would I tell them? That I've hacked every government computer between here and Singapore and something is going to happen Monday." He smiled. "Even I don't have that much credibility. No, we're in this one alone."

CHAPTER 29

Just before first light on Saturday morning Christa drove in silence down Highway 441, turned west on 115 and brought the jeep to park in the small clearing at the foot of Trail #1 opposite the convenience store where the two youngsters had been so brutally murdered. After the murders, Burlingame had closed all his stores from midnight till five a.m. and two clerks could be dimly seen readying the store for the days customers. Anemic fingers of light splayed across the highway highlighting wisps of ashen fog in the soft gray darkness.

Peter had smiled at her as they had gotten into the jeep but had not spoken. In deference to the difficulties he was about to encounter she held her tongue so she would not break his concentration. She watched him step out like a big, silent cat and begin pulling his gear out of the back. The thought of him going into battle as sure as if he was leaving on the crusades, agonized her. He may never come back—this could be the last time I shall see him, she thought. Where are all those marvelous things one is supposed to say at a time like

147

this. That memorable 'going away' line that can be savored forever, and held close to the heart. Words that will bolster him when he's in danger, keep him warm when he's cold, give him courage when he's afraid,—words that will make this all worthwhile. She walked resolutely around the jeep to where Peter was a darker shadow moving among stationary shadows.

"Peter."

"Yes?" His disembodied voice came to her through the mist as though he was miles away. She stepped closer to him. He now had his pack on and was strapping equipment around his waist and on his arms.

Her throat tightened and the best she could do was "Please be careful out there."

He looked at her in the darkness, trying to judge her feelings. His tone was ironic. "Sometimes I get really confused when I'm around you, Christa. When I left here six months ago you didn't appear to be concerned about me—nor us. Yesterday in the kitchen I thought we discovered the beginning of something special—if even for just a minute. Since then you've been—well distant. What's with you?"

She pulled her sweater closer around her shoulders to ward off the early morning chill but the sweater did not warm her words. His attack put her on the defensive. "Please don't blow that out of perspective. You made a pass at me and I just wasn't ready, that's all.

The minute the words came out she would have snatched them back, but too late she remembered an anonymous old Arab proverb—

"…Four things come not back:

The spoken word;

The sped Arrow;

The past life;

The neglected opportunity…"

Peter turned away and adjusted velcro straps around his right wrist. Over his shoulder he snorted, "Considering your background I'd think you wouldn't have that much trouble getting ready."

"You mean once a whore, always a whore, right?" she snapped.

He grinned in the darkness and continued shuffling his gear. At least she wasn't Miss Calm, Cool and Collected as she usually was. She went on bitterly. "I laid on my back for money out of necessity not love. A trade-off. All they got was my body, nobody ever got my mind. They got relief, I got money…"

"Do you still want money?" Peter asked.

"No."

"So what are you after, Miss Christa." He mocked.

Her voice slipped out and hung in the mist like a child's plaint. "I need love—and trust—and compassion, Peter. Three feelings that you have not offered me."

Peter did not answer immediately but buckled the web belt around his waist. He strapped the scanner around his head, inserted the ear plug, molded the voice tube to his cheek and toggled the switch. He slipped the hood on over the scanner and adjusted it for maximum vision. In full gear he looked like a large Ninja. He took Christa's hands in his, his voice muffled a bit by the headpiece. He didn't want to leave with anger between them. "Maybe I'll find those things up there somewhere, Christa. In the meantime let's just say I like you very much."

Christa squeezed his big hands and stepped back. "I like you too Peter, is everything working?"

"Let's see." He switched the chronograph screen on and turned the dial to verbal. Cray droned into his ear. "One target at three o'clock, distance two meters. Two targets at 6 o'clock, distance ninety five meters." Peter dialed 'visual' and the screen on his wrist came alive with three red pulsing dots showing Christa and the two clerks in the store. "Cray?"

"Yes, Mr. Chaney."

"Give me the weight of those targets."

"Target at 3 o'clock 72 kilos, one target at 6 o'clock 61 kilos, one target 6 o'clock 86 kilos."

He grinned at Christa. "You're holding your weight nicely, Ms. Christa."

"I see I'll have to tell Cray what information is classified and what is not when I get back to L'Aerie." Christa laughed. "Now you'd best go, the sky is getting light."

"Yes," said Peter and stepped toward the woods.

"God speed, Peter." she said, and he was gone.

* * * * * *

It took a few minutes to adjust his stride to the terrain and to acclimate his leg muscles to the peculiar rhythm of the Striders but then he flew through the woods like a giant green wraith. The gloom kept him from running full speed but with each long step the trail was clearer. The rifle and pistol velcroed tightly to his legs on either side moved with him. As extra insurance he carried three professional throwing knives and a leather handled survival knife. In his front pack were forty pounds of sensors and in his back pack twenty-five pounds of other, very special, high tech gear.

Inside the first one hundred yards he paused and slapped a sensor onto the bark of a large pine. He stepped away from the tree. The sensor became invisible.

"Cray?" he asked quietly.

"Yes, Mr. Chaney." Cray was loud and clear in his right ear. My God, that's Christa's voice. "Christa, is that you?"

"No Mr. Chaney this is Cray."

"Why are you using Christa's voice?"

"I speak in many voices. This is the one Ms. Christa chose for you."

Peter chuckled behind his mask. When he had first heard Cray back in the parking lot he had known the voice was feminine but had not recognized that as Christa's. Here in the silence of the woods her voice was soft and clear. "O.K. Cray, confirm the signal from this first sensor."

"Confirmed."

"Cray, describe any target registering on this sensor?"

"Two targets. One at three meters. Biped, one hundred kilos. Target number two, eighty meters, quadruped, ten kilos."

Peter switched his wrist scanner to probe and scanned the forest to his back. In just a moment a blip appeared. He switched to full focus and pinpointed the target. It was not visible to him except on screen. He studied it closely. Looks like a fox digging for a rabbit He thought.

"Cray, identify the targets?"

"According to your GPS position the larger target should be you. Target Number two appears to be a small friendly mammal."

"Cray, explain the word, friendly."

Cray answered patiently. "My definition of 'friendly' is any target that I deduce will not harm you. I have been programmed to inform you of all targets but to detail only those which could be unfriendly."

"Thank you, Cray. Confirm each sensor as I plant it." Peter said and proceeded up the path. The fading darkness still restricted him to maximum strides of two meters but he jogged silently on, pausing every fifty steps to slap another sensor onto a tree. As he moved away from each 'planting' Cray quietly confirmed its existence.

In five minutes he crossed Flint Creek. Fifteen minutes later he hesitated at the over grown logging road long enough to make sure the way was clear then bounded across and continued north. The sun was just beginning to define the heavy woods well enough to allow him to lengthen his stride to three meters. He started planting a sensor about every thirty steps.

He began to heat up under the hood so stuffed it into his chest pack. He rubbed a bit of charcoal on his cheeks to break the planes of his face and resumed his jogging.

* * * * *

"Mr. A."

"Yes, Cray?"

"The messages from Ting Sao in Tokyo are being beamed to Singapore and from Singapore to P'yongyang, North Korea."

Jason snapped his fingers in glee. "Gotcha! Cray, can you tell me the destination in the city?"

"The building that houses the National Defense Commission of the Central People's Committee."

"Damn! So we now have access into the heart of the NKPA. Cray, how often are they changing codes now?"

"Each message is encrypted differently, Mr. A."

They are really worried about us now. They can't find us, they don't know who we are but they are sophisticated enough to realize that someone is eavesdropping. "Cray, patch me through to Peter."

"Yes, Mr. A."

"Peter, do you copy and can you talk?"

"Yes to both, Mr. A."

Jason looked up at the wall map. "If our signals are correct you are between the head of Trail #1 and Trail #2."

"Yes."

"You are doing nicely, it is only 7:30 a.m."

Peter laughed. "It's all in the stride, Mr. A. I'm glad you called—I've just discovered something. I'm going to scan it now."

"Cray!"

"Yes, Mr. A."

"Call Christa and put Mr. Chaney's scanner on screen."

The big screen came alive with a haze of reds and grays.

"Cray."

"Yes, Mr. A."

"Translate Mr. Chaney's transmission to real colors please." The screen jumped from reds and grays to reds, greens, blues and whites. "Peter, Cray, focus please." Jason said impatiently.

Christa hurried into the CPU. "What's up Jason."

Fuzz became lines, lines became printing and colors blended into design as the screen came into focus. Jason indicated the screen. "Well I'll be damned, Christa. It's a label from a can of water chestnuts…"

"The WoodChopper…" Christa exclaimed.

"Peter."

"Yes, Mr. A."

"Is this all you've found there?"

"Yes."

"Lay it on the ground, stand up and focus your wrist scanner on it and hold for five seconds. Then turn it over and repeat the process. Cray! video record this."

"Yes, Mr. A."

The image on the screen went to fuzz again then suddenly jumped into focus. It was the front side of the label. It held for five seconds then Peter's arm blocked the view momentarily and then they were looking at the blank backside of the label.

"Cray."

"Yes, Mr. A."

"Give me an *Infrared Spectrophotometer*[9] reading. Look for any fingerprints. How soon?"

"In ten seconds, Mr. A."

Jason drummed his fingers on the table while watching the clock. In nine and one half seconds Cray said. "Mr. A."

"Yes, Cray."

"No recognizable fingerprints. Three round smudges ranging in size from 12.7mm to 31.8mm."

"Cray, compare those smudges with our file records from the FBI's convenience store smudges. Give me a percentage of comparability."

Again Jason drummed his fingers impatiently on the table. It was only five seconds but seemed forever before Cray spoke. "Mr. A."

"Yes."

"On the smudges at the Convenience Store and those on Mr. Chaney's scan there is a ninety four point seven four five six percent probability of match."

"That's close enough for government work," Jason muttered. "Well Peter, you've crossed his trail, put that label in a plastic bag and secure it for me and please resume. We need those sensors in place."

"Right, Mr. A."

CHAPTER 30

"Mr. Chaney! Stop now!."

Peter was running silently down a long sloping hill several hundred meters from where he had found the label. He was about two hundred meters from Trail #2. It took him three giant steps to come to a complete halt.

"Explain, Cray."

"Four small bipeds at one hundred meters, twelve o'clock your present heading. Average weight sixty three kilos."

He felt the hair on the back of his neck stiffen in the primal lust always present before an encounter. He said calmly, "Thank you, Cray."

"You're welcome, Mr. Chaney." He almost said thank you Christa, the voice imitation was so real it was uncanny. He dropped to his haunches and scanned the slope below him with infrared. The four probables were in a group in a small clump of trees. They were not visible but if Cray said they were there, they were there. He crept forward and faded to his left so he could see them. Suddenly a small head came into view eighty meters down the hill. It was wearing an odd looking helmet and was talking quietly to someone else. Without Cray's warning he would have run right into them. He rummaged through his pack and extracted a small, powerful shotgun mike. He connected it to the wrist scanner and pointed it at the head. "Cray."

"Yes, Mr. Chaney."

"Are you hearing that conversation well enough to translate?"

"Yes, Mr. Chaney."

"Good. Give me a literal synopsis translation at one minute intervals. What language are they speaking."

"A dialect peculiar to North Korea, Mr. Chaney."

"Are Mr. A. and Miss Christa aware of this development?"

"Yes."

In ten seconds Cray began speaking quietly into Peter's ear. "Three enlisted men and a Sergeant, Mr. Chaney They have been looking for G-101 for several days with no trace. They have orders to bring him back dead or alive."

Peter spoke quietly into his voice tube. "Mr. A?"

"Yes Peter?"

"I think I aught to talk to these people and see what I can find out."

"As you wish Peter."

Peter smiled and drew a deep breath of anticipation. Here's my first chance to even the score and eliminate part of the threat. These people are blood brothers of that butcher, The WoodChopper. He dropped his pack and other gear and adjusted his shoes for normal travel. He pulled the grotesque pistol from its holster and floated down the slope—a silent green ghost. He stopped when he was but one tree distant from the squad. Two of the North Korean soldiers were lying back against tree trunks engaged in muted conversation. The third was squatting, reading a letter. The fourth, a stern faced non-com stood aloof, at the end of the small clearing, smoking and staring into the woods—his back to the group.

In less time than the telling Peter dispatched the three soldiers with one silent shot a piece. It was the sound of shooting ripe melons in the field. Thump, thump, thump and it was done and he grinned with satisfaction. The non-com turned curiously at the sound and saw the squatting soldier slumping backwards to the ground. The startled

Sergeant strode into the center of the clearing and his eyes widened with the shock of seeing blood pouring from the single tiny hole in each of the soldier's heads. At that moment Peter stepped silently from the fringes of the pine tree, emerging like a monstrous green golem pointing a very long, dangerous looking weapon at the officer. Surprise, fright, incredulity struck him in a bundle and he paused for one fatal heartbeat. In one long stride Peter caught him just below the left ear with a rigid, slashing chop. The Non-Com folded into a ragged heap like a spider hit with insect spray.

Peter pulled his arms behind him.

"Mr. "A"—I'm tying up the Sergeant now. He's in pretty bad shape but I'll try to bring him around and talk to him. The conversation may not be pleasant—you and Christa probably should not listen."

"What about the other three, Peter?"

"They have been,—neutralized."

Christa watched Jason wince but he answered quite calmly. "Proceed as necessary Peter, we desperately need information."

* * * * *

The pain in his jaw was a searing, jagged evil that ranged to the top of his head and down the side of his neck. The pain itself almost brought him to his senses. But somebody slapped him, not gently, on the ear and the evil raged across his tongue and cramped his throat. He tasted blood in the back of his throat and swallowed to keep from choking. He opened his eyes slowly and tried to move his jaw. The effort caused excruciating pain that swelled into his head and turned everything fuzzy. Something moved in front of him and he focused on it with great effort. It was the giant green golem and he was speaking.

"Cray?"

"Yes, Mr. Chaney."

"If this Korean speaks Korean I want you to translate and relay back to me verbally."

"Yes, Mr. Chaney."

"Do you understand English, Sergeant?"

Spit mixed with blood was drooling from the side of his mouth and he tried to wipe it off but discovered his hands were bound behind his back. Mustering all his courage he glared back at his captor. Dark, brooding eyes regarded him through the eye holes of the green hood.

He felt like a wounded sparrow confronting a King Snake. The golem reached down and tapped him on the left side of his jaw and splinters of pain drew a red mist across his eyes. He groaned in renewed agony.

"I repeat, Sergeant, do you speak English?"

He managed a grunt. "Uh."

The green Golem put the needle sharp point of a knife against the tip of the Sergeant's nose and rasped.

"Is that a yes, or a no, Sergeant?"

"Yiss," he hissed.

"I need information and I don't have a whole lot of time to wait for the answers. What is going on back at your base?"

His jaw was on fire and the tip of the knife was biting into his nose. He mumbled through bloody froth. "An enlisted man deserted. I search for the dog."

Peter put more pressure on the knife to emphasize his impatience. Blood welled up around the point and dribbled into the corner of the Korean's mouth. Agonized tears formed in the dark, almond shaped eyes. Peter said. "I know all that, I want to know what's really going on back at your base?"

With great effort the Korean said, "We do cancer research…"

Peter exploded! "Bullshit! You don't need a cadre of elite soldiers to guard a cancer research center…" and he punched the knife down into the Lieutenant's nose and pulled it out sideways.

Added to the broken jaw the pain of the split nose overloaded the soldier's senses and he blacked out. Peter looked down at him with disgust. A gout of red blood gushed from the Sergeant's nose and bubbled into his mouth. Instantly he began to cough and to choke, then to kick violently and suddenly he lay back, writhed and was still. He had drowned in his own blood.

"Shit, shit, shit!" moaned Peter.

"Cray?"

"Yes, Mr. Chaney."

"Connect me to Mr. "A.""

Jason came on-line. "Yes Peter?"

"The Sergeant has joined his men. He gave me no information we didn't have already."

"Unfortunate. Are you all right?"

"Yes."

"Then continue with the sensors."

CHAPTER 31

"Christa?"

"Yes Cray."

"The survey of satellite geodetic records for the past eight years is concluded."

"Did you find any activity that fit the criteria?"

"No, there was no traffic in daylight."

Christa turned to Jason. "Evidently everything for the construction was moved in during the night."

"I'm not surprised. That's what I would have done. Cray? Were there any infrared scans?"

"No Mr. "A"."

"Well," mused Christa. "we've come up dry on the overflies, the final plat of Voodoo Mountain was never filed and there are no architect's plans available, do you want to use a Condor?"

"No. The satellite photos plus the Air force overflies should give us the resolution necessary to pick the specks out of the pepper. Besides, I don't want to alert them to that level of scrutiny yet."

* * * * * *

"Commander."

Colonel Pong-Yul toggled a switch on his desk radio. "Yes, Lieutenant?"

"Sir, Squad Number One is thirty minutes late reporting in on schedule."

"Can you not raise them at all?"

"Negative."

"Keep trying and let me know of any change. In the meantime alert the other squads."

"Yes, Colonel Sir. Also, Sir, a message in your code is coming through from headquarters."

"Send it to me."

"Yes, Colonel Sir."

Within minutes there was a knock at his door and the enlisted messenger handed him a sheet of paper. When decoded the words sent a chill down his spine. It said simply. "Unknown powers are attaching trailer to all your messages. As of this moment reduce traffic to irreducible minimum and random switch encryption each message. Transmit to Destination B. B will re-encode and transmit to headquarters. You are responsible this project. If this development effects outcome it will likewise effect your future. End."

He looked at the word 'end' for several seconds than spoke into the radio.

"Watch Commander."

"Yes, Colonel Sir."

"As of now change encryption for each message on a random basis and transmit to Destination B. Send only those messages voice authorized by me. Understood?"

"Yes, Colonel Sir."

One of the disadvantages of living underground is there are no windows to look out, no vistas to salve the mind, nothing but four walls broken only by drab doors leading into drab halls. A fair hand with brushes and colors Colonel Pong-Yul had used his off time painting a six by twelve foot mural in light, bright colors on one wall of his living quarters. A small waterfall at the left side led into a pool of clear, sparkling water full of brilliantly colored fish surrounded by grassy banks, stately, gnarled trees, and large moss covered rocks. He turned to it now. He could almost hear the waterfall, rolling and tumbling in its rush to join the calm, peaceful water below. In his minds eye he wished he could step into that picture, sink down upon that grassy bank, drop his bare feet into the water and let it wash his troubles away.

There are forces at work in the woods that I do not know about. Something, or someone has injected an unknown equation into my carefully planned formula and it could distort the whole outcome of this project. Damn G-101. The problem is, there is some one else out there and I don't know who. But all I have to do is hold this thing together for two more days and then it won't make any difference what the foreign dogs do. But why didn't the Supreme Council do all this experimenting and preparation some where else and simply bring the subjects to the United States when the project was ready. It was an insoluble problem.

He sat on the edge of his bed, laid his mind down on the grassy knoll alongside the peaceful water and filled his head with memories of his family.

In a few short days this ordeal will be over and I will go home to a hero's welcome—my family and my world will welcome me with open arms. No longer will I have to endure the slurs and slander—my deeds will have purged me of the stigma of foreign blood.

He slept.

CHAPTER 32

Peter moved carefully up a small draw and over the ridge to intersect with the northern end of Trail #2. It was much slower planting sensors when running south than north as he had to nearly stop to 'plant' the sensor. During one of these pauses he caught a glint off in the woods. Cray was silent so he went to it. Twenty feet off the trail he kicked at two vegetable cans originally buried in the heavy pine needles but since uncovered by curious squirrels. His heart leaped as an insect repellent can came to light.

He went through the drill with Cray doing an infrared spectrophotometer. There were smudges present and they matched the other two samples. Cray estimated the age at less than seven days.

With all his senses on alert he continued jogging south on Trail #2. Without incident he reached Flint Creek and stepped off the trail a few yards to enjoy an MRE for lunch. Sweet and sour pork on rice was a nutritious choice for lunch. He washed it down with an orange drink and topped it off with a piece of lemon pound cake and one of Christa's energy tablets. Benevolent high noon sunlight filtering through the pines above him lulled him and for a few moments he lay back to bask.

"Mr. Chaney."

Christa's voice came to him from far away. Reluctantly he answered. "Yes Cray."

"Mr. "A" wants me to read you this police report. quote: The summer home of Tom and Marie Kasman at the extreme north end of Lake Pawnoo was broken into sometime this last week. The burglar left with assorted canned goods, other incidentals and a 243 Remington Varmint rifle the Kasman's kept for home security. No damage was done to the cabin...unquote."

"Peter?"

"Yes, Mr. "A"."

"Looks like our boy is up there in the north between the lake and Voodoo Mountain."

"Did this go out over the local radio station?"

"Yes, so Voodoo Mountain will also be aware of the break-in."

"Yeah," said Peter.

"Peter, I want you to retrace your steps up Trail 2 till you reach The WoodChopper's trash. Then cut straight east to Trail #3. Plant sensors south on three until you intersect with Trail 4 and plant northeast on four along the backside of the lake to the Kasman's house."

"You think he may still be in the vicinity?"

"Probably not close but you may pick up some kind of trail from there. Cray has been working on the probabilities and has placed The WoodChopper most likely in the area of The Devil's Play Pen."

"Why not go directly there?"

"Might work out all right in the long run but we need the security those sensors give us in the meantime. How are you physically?"

"I've never been better."

"Good, then go." said Jason and clicked out.

Peter buried his debris, turned his boots to maximum and pointed himself north back along the way he had come. Running with the boots at max gave him a stride at times up to five meters and required his full concentration each time he came down to compensate for the unaccustomed thrust. Matching his body to the giant steps was an exhilarating experience and ended too soon with the sighting of the spot where he had discovered The WoodChopper's cache. He readjusted the boots and moved silently off through the woods to his right in search of Trail #3.

"Mr. Chaney, stop now."

"What is it Cray?"

"Two probables at ten o'clock, two hundred fifty meters, and four probables at twelve o'clock, two hundred seventy two meters."

"Well Cray, I was beginning to think I was out here all by myself."

"You have not been alone since you started, Mr. Chaney."

"Really, Cray. Explain."

"I have twenty-two separate sightings of friendly targets including the one standing fifty meters behind you now."

Peter shivered and turned quickly but saw nothing. "What is back there, Cray?"

"It is four legged and weighs 56.8 kilos." White tail deer, thought Peter. "Cray, give me directions on the probables."

"The group of two are traveling south on Trail #3 and appear to be hikers. At present speed they will cross in front of you in three minutes, forty two seconds. The second group is receding in single file. At present speed they will pass out of range in two minutes."

"Cray, get me Mr. "A"."

"Go ahead Peter."

"The hikers are no problem, but I think I aught to pursue the squad."

"No, Peter. We need those sensors down three and up four so I want you to follow the plan. If all goes well you'll be around the top of the lake before they will be anyway."

"O.K." Peter said reluctantly. "I'll verify the hikers and then go around them and finish up."

"Peter, it might be a good idea to escort them out of the woods or at least make sure they are safely on their way out. They must have gone up Trail 3 early this morning. If they stumble across a squad they're dead meat."

"That's really going to slow me down."

"I know," said Jason. "but it's the only safe way. All hell is going to break loose up there in the next twenty fours hours and we need to know they're out of harm's way."

* * * * * *

Peter let the two hikers drop below him until they were just out of earshot then stepped onto the path and followed at a slow jog. He turned on his wrist scanner and kept his distance—the last thing he wanted to do was spend part of the afternoon explaining his appearance to two freaked out hikers. He paused every ninety steps to slap a sensor against the southeast side of a big pine. When the hikers reached the junction of Trail 3 and 4 he breathed a silent sigh of relief when they ignored 4 and continued on their way toward the highway. He turned left, started up 4 and almost immediately began to climb with an occasional glimpse of the lake off to his right. Soon he was forty feet above the calm water, looking down the vertical cliffs that bordered the northwest shore. The path wound above the cliffs around out croppings of rock, through pine thickets, over fallen trees and circled berry patches. In one spot Cray routed him away from one berry patch in which he heard a large brown bear foraging for his afternoon snack.

He rounded the northwest end of the lake and began the descent. "Mr. Chaney, stop now."

Peter stopped. "Yes Cray."

"There is one small probable northwest of you and receding. Target is at my extreme range—and gone. Please sweep the area now with your wrist scanner."

Peter snapped the scanner on and made a slow arc to his left. "Stop!" said Cray. "Sweep five degrees slowly, there! Hold there!"

Peter stopped and looked at the screen on his chronograph. A red blip appeared and moved slowly across the screen. It faded in and out.

Cray said. "Target is now at two hundred meters and fading. Much interference between, am now at maximum scan.

"Thank you Cray. What does it weigh?"

"Sixty four kilos."

Hmm, Peter thought as he watched the blip drop off the corner of the screen then continued on down around the lake. One hundred forty pounds, about the right size for The WoodChopper. He was sorely tempted to give chase but did not.

Locating the burgled house was no problem. It was the first one you came to on the north end. He saw the boat dock through the trees first and skirted around to the left using the trees for cover.

A four wheel drive station wagon was parked in the quiet yard. Late afternoon sun slanted through the trees and highlighted a contented squirrel picking up a nut from the ground. Peter focused the wrist scanner on the house and turned it to max. "Cray, can you tell me who is in the building?"

"Mr. Chaney, sweep the building slowly." Peter did and Cray said. "Two probables, one sixty kilos, one eighty four kilos. Two friendlies, one two legged, forty kilos; one four legged, twelve kilos."

"Thank you Cray." The Kasman's, thought Peter. Mr. and Mrs., the child and the dog. He turned the wrist scanner off, drew an imaginary line from the back door northwest into the woods and concentrated in a grid search of it. Nothing showed up and dusk caught up with him forcing him to give up. Even night goggles were not good enough to spot the kind of sign he wanted.

He withdrew five hundred yards into the woods to the northwest of the house and selected a spot where a large oak spanned the faint trail. He relieved himself, climbed fifteen feet into the leafy arms of the giant and strung his hammock. As the moon filtered through the woods he climbed into the hammock, ate a leisurely MRE, drank half a pint of water, bid Jason and Christa good night and asked Cray to wake him at 4 a.m.

CHAPTER 33

Sleep did not come easy this night. Usually he slept the sleep of the dead but tonight the proximity of the lovely Christa intruded on his peace and he dozed fitfully. He could still hear her say, "I need love—and trust—and compassion..." I cannot give you those emotions, He thought. I'm like an android, without them in my makeup. Love to me is a physical expression, an act, you do it with someone else. I don't trust anyone but myself, well—maybe I have to include Mr. "A" in that, but certainly no one else. And compassion, compassion is a luxury reserved for fools, and I am certainly no fool. But I would dearly love to make love to you, my lovely Christa....

Cray woke him quietly at 3:30 a.m. "Peter, there are four probables just coming into range Northwest. They are closing rapidly. Current heading will put them directly underneath your position in three minutes."

Peter slid out of the hammock, sat on a large branch and swept the area above him with his wrist scanner on maximum. Four blips appeared and slowly traversed the screen. They were close enough for him to hear an occasional mumble through the leaves but were still too far away to pick up conversation with the shotgun mike. He tensed as they broke through a small clearing just up the hill then suddenly they were walking silently, single file beneath him, just four darker shadows slipping through the woods. They halted fifty yards down hill.

Peter came down the tree like a stalking cat and moved toward the squad. The path, unlike the woods, was relatively free of pine needles and leaves, making it possible to walk slowly and silently to within ten feet of the group. The soft lilt of the North Korean language being spoken in a whisper filtered through the trees to his mike. Cray began to translate the essence of the conversation. They were looking for The WoodChopper with orders to return him or kill him. The consensus was that there was more glory in killing him. One of them detached himself from the group to take a pee and walked straight toward Peter. Peter slid the mike into his jacket and slid the wicked, dark bladed survival knife from its scabbard—and waited. The soldier was no taller than Peter's armpit. He stopped three feet in front of Peter and began unbuttoning his fly. Peter stepped forward, caught him round the throat with his left hand and punctured him just below the sternum with an upward thrust of the needle sharp blade. The small form sighed in the darkness, then went limp under his hand. Peter lowered him softly to the ground and wiped the blade on his tunic. That must be what it feels like to kill a child, thought Peter.

He stepped quietly away from the body and waited. It wasn't long before Cray began to transmit the squad's uneasiness and they called softly to their comrade. One form, hunched over, his rifle at the ready, left the group and slid through the woods towards Peter. Peter held his breath as the small form passed him at less than arm's length, then drove his knife deep into the Korean's kidney. He dropped without a cry but Peter missed the rifle in the darkness and it hit the ground with a clatter.

Peter stood still and waited. Cray picked up an excited flurry of words coming from the clearing. The remaining two were the radio man and the Sergeant in Charge. The Sergeant was whispering instructions and the radio man was calling Voodoo Mountain.

Cray placed the two in the middle of the clearing apparently standing back to back for safety's sake. Peter slid the Anschutz from its holster and moved silently in a circle through the trees to flank the

two men. He came to an open spot in the brush and could just make

out the shadows of the two standing, back to back, locked in fear in

the middle of the small clearing. He raised the Anschutz, aimed the

laser dot slightly to their right just above where their shoulders aught

to be. He panned to the left in a slow, controlled arc. As the red dot

crossed the radioman's neck he squeezed off one silent shot. The

copper jacketed slug bored a neat hole through the radio man's

trachea, then through his spinal column and opened a quarter sized

hole in the back of his neck. From there it reversed the process with

his Sergeant who was standing back to back with him. Both men

dropped as one.

Peter aimed his miniature flash at the scene. Both lay where they

fell, blood slowly oozing into the pine needles. The radio crackled

with a muted question. Peter inspected it closely then reached down

and flicked the switch to receive.

CHAPTER 34

"Commander!"

The sleeper fumbled in the darkness, snapped on a bedside light.
"Yes?"

"Sorry to bother you, Colonel Sir…"

"Get on with it!"

"I lost Squad 2 right in the middle of a sentence."

"Explain!" He snapped. He looked at his watch, lit a cigarette,
took a long pull as the frightened but calm voice came back into his
room. "Sir, per your orders. Squad Two was sectored in a small
clearing three hundred meters short of the house on the north side of
the lake. They were waiting for first light. Something, or someone,
incapacitated two of them in the dark and the other two were standing
back to back for maximum security. I was talking to the Leader when
he suddenly made a choking sound and went silent…"

The Commander sat up, put his legs on the floor and started to
dress. "And?"

The disembodied voice of the Radio Operator came crisply through the bedside speaker. "Nothing Sir. When he stopped talking the switch was still open. I heard him fall, I heard rustling in the bushes and then someone switched the radio to receive."

The Commander was buttoning his tunic. "Have you heard from Squad One?"

"Negative, Sir."

"Keep listening, inform Number Two of this and tell him to meet me down there in fifteen minutes."

"Yes, Sir."

CHAPTER 35

In fifteen minutes first light came cautiously through the trees turning black shadows to gray, then to lighter gray and finally gave the forest color. Peter surveyed the scene. He dragged the two outlying bodies into the clearing and searched them. They had no money, no coins, no pictures. Obviously they had been ordered to carry no identifying material with them. The radio kept up an insistent nonsense that Cray interpreted as nonessential and Peter ignored. He pulled the bodies into a small copse, collected the Korean radio and his own gear and resumed the search for The WoodChopper.

As he tacked up the side of the mountain the trees changed from the dark green of the pine needles and took on the many leaved, light green of the hardwood. Oaks, hemlock and hickory vied for the sunlight, turning the forest floor into a cool thicket of bush and briar. The density made the tracking easier because the deer took the path of least resistance and their trails were plainly marked. A mile up the slope the terrain changed again to rugged short peaks, steep small valleys, scrubby pines and volcanic rock.

A bit of bright in the pine needles caught his eye. He stopped and nudged it with his toe. A small, clear plastic, water bottle. He scanned it and felt the old familiar rush of adrenaline when Cray identified the now familiar smudges, judging them less than twenty four hours old.

He moved cautiously on up through the sparse growth, threading his way around the hillocks until he reached a vantage point above the barren Devil's Play Pen. The rotten odor pinched his nose. He resisted the urge to cough and spit, knowing the human sound would carry deep into the Play Pen. A silent hell, over which no bird flew nor animal trod, no sound came from the sterile landscape stretching below him. Even the fumaroles puffed and belched and fouled the air in silence. Here and there giant hillocks, like African anthills stood like cone shaped warts, their tops glistening in the sun. Inside some were eerie caves, carved by unknown forces, which were now the object of his scrutiny.

He turned his scanners to maximum and slowly traversed the area but came up with nothing. He was using his binoculars when Cray's dulcet contralto broke the silence. "Mr. Chaney."

Peter flinched at the voice. It was most difficult for him to remember that this was an electronic apparatus, not Christa talking to him. The hair was standing erect at the nape of his neck again. "Yes, Cray."

"Four probables have just entered sensor range on Trail 3, traveling south."

"Cray, why do you list them as probables?"

"There are no civilians registered in the area. They are moving at a fast walk. They are in squad formation."

"Thank you, Cray."

"You are welcome, Peter. They have now departed Trail 3 and are fading out of sensor range heading due east."

"Cray, probable destination?"

"The Devil's Play Pen, Peter." Cray said, her voice matter-of-fact.

"Cray, please ask Mr. "A" to launch a Condor for added surveillance over the Play Pen."

Her voice came back almost immediately. "Peter, a Condor will be launched in three minutes."

"Thank you Cray."

Peter busied himself among the rocks setting up a vantage point that could not be easily observed. From his pack he pulled a tubular package about the size of the card board tube in a box of plastic wrap. It looked similar but there the similarity ended. It was a Photo Refractive Polymer Wafer. He slid it under his camos to pick up some heat from his chest. As it neared body temperature he could feel it begin to expand. He took it out, and unrolled it gently across his lap. It flattened into a one half inch by six inch by six inch tablet, one side smooth as a mirror. On the side of one corner he plugged in a cord and connected the other end to his Chronograph. He pushed a button on the Chronograph and the six inch screen glowed and came to life with a moving picture. As the picture continued to change he realized it was being transmitted from the Condor (which was on its aerial way to the Play Pen) to the satellite, then back to Cray and back to the Satellite and back to his Chronograph and thence to the wafer. Seemingly a complicated way to get information but he knew in that configuration Cray could analyze every bit of data coming off the Condor scan.

He inserted his miniature halogen flashlight into a round slot in the side of the wafer, snapped it into place, turned it on and the picture on the screen morphed into three dimensions. The Condor was now hovering on wide scan at one thousand feet above the Devil's Play Pen but he could neither see it nor hear it.

"Peter."

"Hey, Mr. "A"."

"I'm going to grid your screen and you'll see where you are in relation to the GPS. Every thing else all right?"

"Yes."

"Standby."

Peter's screen blinked and when it rose back into focus grid marks popped up like a tic-tac-toe game screen. A pulsating blip caught his attention immediately.

"Mr. "A" What about this blip?"

"That's you Peter," Jason said reassuringly. "Now I'm going to high resolution infrared scanner on the Condor."

The screen changed color, lost much of its acute focus but immediately showed four blips entering scan range on the west side.

Jason came back on-line. "Now these are our four probables that Cray reported coming in off Trail 3. They've just crossed the old logging road coming east and they're a bit more than two thousand meters away from you. According to Cray, if they continue on course and speed they'll be on top of you in twenty minutes give or take."

Peter grinned and shucked his hood. "I'm ready for them. Incidentally, where's all the wildlife on the infrared?"

Jason answered, "Cray is filtering it out. Anything under forty-five kilos doesn't show and only above that if she can determine bipedal gait."

Very smooth, Peter thought. Man is the only animal on this earth that walks any distance on two legs.

Jason was looking at another screen as he went on. "Actually the woods below you is swarming with small game and even a bear or two that show up on the other configuration."

"Keep me posted."

"I will, whoop! there's another blip right out in the middle of the Pen. Do you see it?"

"Yeah, looks like about eight hundred meters down there. Let me use the binoculars on it."

Peter threw the glasses to his eyes and scanned the terrain below. Whatever the blip was it was not yet visible except to infrared.

"Peter, I've got it on TV scan now. Look!"

The screen blinked and the overhead view of the Play Pen came back into focus. Jason was zooming into close focus and a human figure popped into view. It was a small man wearing a red and black jacket and a baseball cap—The WoodChopper. He was scrabbling for cover. He dropped behind one of the hillocks and put a rifle to his shoulder. The sound of a shot wafted up to Peter. Jason said. "One of the four soldiers just went down. Now they're looking for him."

The screen shifted back to wide view and Peter could see the three little ants as they worked their way over the tortuous terrain generally towards The WoodChopper. There was another popping sound from The WoodChopper and one of the three remaining soldiers paused, but then came on with the other two. There was now only one hundred and fifty meters between them and their quarry. Two of the soldiers dropped to firing position and a fusillade of shots rang through the warm air. Unharmed but spooked, The WoodChopper broke cover and started scrambling up the path out of the Play Pen. His flight took him directly toward Peter's vantage point. One soldier dropped farther and farther behind and finally fell down but the two remaining doggedly closed the distance between them and The WoodChopper.

The WoodChopper in full panic stricken flight scratched his way up the rugged path toward Peter. But he was a hundred meters down the hill and it was obvious he would fall into the hands of the two remaining soldiers. Peter unlimbered the Anschutz, sighted it over a rock and got his first silent shot at the front soldier. The man went down in an awkward sprawl. The one remaining soldier, stepped across his fallen comrade and wound his way through the rocks and up the hill after The WoodChopper.

Peter put the pistol down and stepped around the rock to confront the gasping WoodChopper.

Both men were startled but The WoodChopper reacted a split second faster. Peter saw a short, muscular, olive skinned monstrosity wearing a baseball cap on a perfectly round head with no hair, no eyebrows, two droopy round holes for eyes, two punctures where his nostrils aught to be, and a slit for a mouth. But he waited too long taking this glimpse and The WoodChopper was on him like a vicious little wolverine. Instinct brought Peter's hands 'on guard' and saved his life but he took three ferocious blows with the last one being to the side of his jaw just below the ear. The world exploded and he went down and out. He fell back among the rocks.

The WoodChopper paused to finish the job but scrambling, heavy breathing noises from down the path signaled the arrival of the pursuing soldier. G-101 grimaced and bolted headlong up the trail. A long straight stretch proved his undoing and a burst from an AK47 cut into his shoulder and he went down. His wiry little pursuer was upon him with a vengeance, clubbing him on the ground. G-101 collapsed and his tormentor dropped wearily beside him piling a stream of invective on his unconscious captive.

* * * * * *

Jason heard the quick intake of Peter's breath, matched the thuds with the action and heard Peter hit the ground. Then there was silence, then shots and muffled Korean in the background. "Cray, translate please." he begged.

"It is all Korean invective, Mr. "A".

"You mean he is cursing, Cray?"

"To use the vernacular, yes, Mr. "A". Quite a stream of it."

"Cray, Is he saying anything about Peter?"

"Not specifically."

"Cray, summarize that statement please."

"The transmission includes the information that he knows someone else is out here hunting G-101, but he is not aware who."

Jason glanced at the screen. "The Condor shows that soldier at least a hundred feet from the last place we saw Peter…"

He looked at a white-faced Christa. "The cat's out of the bag, but if we can raise Peter we should be all right." he said this with far more certainty than he felt. "Christa full zoom on the Condor and lets see if we can see anything."

In this case the Condor was blind. Peter was lying between two rocks and under an overhanging ledge and was concealed from overhead sight.

Jason called Peter several times and then Christa tried with no luck. Cray came back on air. "Mr. "A"."

"Yes, Cray."

"The Korean soldier has radioed his base and has been instructed to make it to Rocky Flats where they will be picked up by a helicopter."

Jason consulted the big wall map. "Yes, it's the only place a chopper can land up there. What's their ETA (estimated time arrival) at the Flats?"

"They are hampered by G-101's wounds so it will be approximately thirty five minutes, Mr. "A"."

Peter, Jason thought. Wake up. We need you.

CHAPTER 36

Peter came to in a red fog with a picture of The WoodChopper floating around in his head. He heard Christa's voice off in the distance. "Peter, talk to me! Peter, please—talk to me."

His first impression was that his head ached, his jaw was throbbing and as he came fully awake his ribs felt like the flesh had been ripped off them where The WoodChopper's vicious blows had landed. That little fella fights for keeps, he thought…

"Peter, answer me, please."

"Christa, are you live or am I talking to Cray?"

"I'm me. Are you hurt?"

Peter stood up slowly and took a couple of steps. "Everything works but I've got some aches and pains I didn't have before."

"What happened?" Jason and Christa asked in unison.

"That little bastard attacked me—and he's good. He caught me by surprise but it won't happen again. Guess what he looked like?"

"What?"

"This guy is a little monster, he's grotesque. He's something out of Dick Tracy. I'd call him round head. A perfectly round head, two holes for eyes, two holes where his nose aught to be, a slit for a month and no hair at all…"

Christa sucked in her breath. Jason asked, "Just like Karen's drawing. How big is he?"

"Probably five foot two, one hundred forty pounds, full of muscle and very fast."

"That agrees with Cray. Peter, can you travel?" Jason asked.

"Yes."

"Good. Collect your gear and make for Rocky Flats post haste. That's where they're taking The WoodChopper. A chopper is to meet them there and ferry them back to Voodoo Valley. Do what you can to sabotage that plan and get The WoodChopper"

"Cray, what is the ETA at Rocky Flats for the probables?" Peter asked.

"Twenty nine minutes, Peter."

Peter began condensing his gear. "Mr. "A", where are they now?"

"West, Southwest of you Peter, We're tracking pretty well with the Condor but the tree cover is so thick we lose them from time to time. Rocky flats is on the Northwest side of the old logging road North of where the road crosses Trail #3. If you go straight West to the road, then turn south you can make the best time. Good luck and keep the mike open."

Peter plucked the last of his gear off the ground and loped off through the woods. Cray gave him a clear field and he traveled without caution. Ten minutes into his trek Cray said. "The two probables are at four hundred eighty meters West, Southwest."

"Cray, what is their ETA now at Rocky Flats?"

"Subjects have increased their pace. ETA is now twelve minutes."

Peter reached the logging road, adjusted his Striders for maximum distance and bounded south. "Cray, what is my ETA at Rocky Flats now?"

"Eleven minutes Peter."

"Peter!"

"Yes, Mr. "A"."

"Their chopper is in the air just clearing the caldera on its way to Rocky flats. It may pass right over the top of you. We missed it coming over the edge and now it's traveling too low for me to scope it so we have no idea who's aboard. We do know they're aware of you somewhere out there so beware."

CHAPTER 37

Lieutenant Chong Sang-Tu loved to fly choppers and his ships always knew it. When he strapped himself into the pilot's chair it was more the ritual of putting on a form fitting glove than sitting down in an iron bird. Once in the seat he became only mind and his body just an instrument to enable him to transmit his thoughts to the waiting bird. He never consciously thought of where he wanted to go in a chopper or of what he wanted it to do, he simply willed it to do his bidding and it did.

He had cleared the caldera and dropped immediately to tree top flight. The sky was full of enemies and the only way to thwart them was to stay low. The Purchasing Committee had made a wise choice in buying American Choppers to use on American soil and the modified Twin Turbine Bell 222B hummed along like a great black shark cruising at one third speed. Bell Helicopter had never intended their sleek ten passenger ship to be anything but a people mover but his masters had seen it in a totally different light. With its twin turbines and light weight it was an excellent choice as a mobile crane.

And fitted with a 23 caliber Gatling Gun and fifteen thousand rounds of ammo slung between the skids its one hundred forty five knots fast cruising speed made it a formidable gun ship. Loaded now with an AK47 four man gun squad Lieutenant Chong Sang-Tu was a buccaneer praying for his moment of glory.

His Co-pilot pointed out the window to port at the covert ribbon of the old logging road dropping away below him. As he was about to cross it he lowered the collective, decreased the throttle, adjusted the cyclic and nudged the right pedal. The sleek ship rolled slightly to the right, came level over the logging road and flowed south with it. He bumped his infrared scanner to full power, reduced his speed to thirty knots and probed the road like a mongoose sniffing out a cobra. It was along this road his prey aught to be traveling if he was trying to make time. With any luck they would cross paths.

* * * * * *

"Peter?"

"Yes, Cray?"

"Your targets have crossed the logging road and are proceeding up the hill to Rocky Flats. ETA five minutes."

"Peter!"

"Yes, Mr. "A"."

"The closest point on the road to Rocky Flats is about two hundred meters below you. According to the map there is a trail leading up to the rock. It goes around the rock and leads you onto the west side because there is a sheer drop here on the east..."

"Wait a minute." Peter broke in. He stopped and tried to listen above his racing pulse. "I have a chopper about three hundred meters north of me."

"That's going to slow you down a bit."

"Not unless they're looking for me. I'm pretty well invisible."

"Their primary orders are to bring G-101 in but remember they know about you."

"Yeah, I've got to go. I can see them through the trees now."

Peter resumed his lengthy strides along the road. The over growth was too thick around the old road to allow him to pick a spot two hundred meters away so he simply had to count steps. The helicopter came round a turn, swooped slowly in behind him and hovered. Like a cat poking around in a trash barrel it seemed to be looking for something but not quite sure where it was.

"Peter," Cray said laconically, "You are being scanned with infrared."

In mid-stride the revelation came. They were scanning him with infrared but with the disguise they were having trouble pinpointing him. He looked back and up at the chopper which was hovering less than a hundred feet above and slightly behind him. His heart turned cold and he hit the ground with both legs driving back against his body stopping his forward movement. Underneath the belly of the chopper was the long tubular shape and the ugly multiple snout of a Gatling gun and it was seeking him. He dodged sideways into the woods and the gun began to chatter.

The logging road exploded where he had been standing. Dirt, dust, rocks and bullets filled the air with stinging particles. They flailed at him as he hit the ground in the pine needles across the ditch and bounced forward four meters. He went down on all fours and scrabbled behind the large boll of a friendly oak tree. The stream of bullets followed him into the woods and he could hear them thunking into the trunk behind which he was cowering. The hail of slugs cut and cleared the brush on both sides of the massive tree and brought some of the lower branches down on the side toward the chopper.

The gun paused as though to reconsider and Peter squirted directly away from the tree keeping it between him and the bird. In a matter of moments he was fifty feet uphill behind another tree when the pilot opened up again and used the hammering gun to saw the upper branches off the tree. They popped and cracked and dropped like shatterings in an ice storm.

Peter readjusted his shoes and moved quickly away from the havoc. He covered a hundred meters before his hearing came close to normal and he could hear Jason calling him.

"Yes, Mr. "A", I'm here—and okay." While he continued up the path he brought Jason and Christa up to date.

Jason answered. "From what you say Peter, they never did see you, they just caught you on infrared. Be careful where you are now because if he follows you up the hill and continues scanning he may find you again—and this time we may not be so lucky…"

"Where's G-101."

"Condor shows them on the top of the rock waiting for the chopper."

"Do you want me to take a shot at the chopper?"

"Yes-s, but keep in mind what you just went through. You've got the firepower to bring it down but if they find you again you may not have any place to hide."

The path approached a perpendicular cliff, abruptly turned left and began to climb up and round the side of the rock. Peter could hear the beat of the rotors above him as the chopper hovered above the huge flat table of the rock. He increased his pace up the path.

* * * * * *

Pilot Lieutenant Sang-Tu held the bird hovering in the air like a leaf defying gravity. He was still chuckling over the interlude below in the woods and even though he did not get a confirmed kill the magic thrill of the hunt was coursing through his veins. To see the tree disintegrate before his eyes was a thing to behold. He had left the giant oak a shambles of twisted limbs punctuated in the middle by a stark pock marked trunk. Sang-Tu could still hear the staccato bark of the gun and feel the warmth of it through the floor of the bird. What a day! He banked the chopper ever so slightly and slid it down toward the two figures on the rock. The prop wash buffeted the two and the wounded one fell down. The other jerked him roughly back to his feet and they waited with heads bowed.

Sang-Tu sat the bird down gently and the squad poured out to form a half perimeter, guns pointing in four different directions. The soldier forced his prisoner forward, and both of them, half blinded by the prop wash, stumbled toward the ship.

The Co-pilot tapped Sang-Tu on the arm and pointed to the infrared scanner. A red blip was moving slowly across their right front. He looked out the window. There was a waist high cluster of rocks on the west side of the flats and as he watched he saw a large green shadow flit from one rock to another. He leaned out the door, screaming at the squad leader and pointed at the rocks. As a single man the squad opened fire and scourged the rocks with a lethal bath of lead.

G-101's captor pushed him to the door of the chopper, threw him aboard and jumped in after him. One by one the firing soldiers ceased fire and jumped aboard. When the last was in and the door was being secured, Sang-Tu pulled the anxious bird two feet into the air then slid it back over the precipice and dropped it away from the surface of the rock.

CHAPTER 38

"Peter, are you all right?"

Peter was wrapping green gauze around his arm. "I've got a few small nicks but nothing serious. Man they laid it down."

Jason laughed. "Thank heaven for favors. We watched the whole thing from the Condor. I couldn't help because that condor has nothing but photo gear on it."

"Where's the bird now?"

"Well on its way back to the caldera. Christa is following with the Condor and that's where you're heading next."

"I'm on my way."

"Have you still got some sensors?"

"Yes."

"Good. Plant some as you go up the side of the mountain for insurance behind your back," Jason paused for a moment. "Just getting the info from the Condor. The chopper disappeared into the side of the mountain and everyone is underground. I'm afraid you're going to have to follow them."

* * * * * *

Colonel Pong-Yul strode purposefully along the corridor, knocked on Dr. Ho-Sik's office door then entered without waiting for permission.

The tiny Doctor turned an impassive face to the intruder, his eyes owlish and unreadable behind round glasses. The room was unlit except for a shrouded desk lamp which pushed the darkness just beyond the edge of the desk. The lamp defined only the lower half of the doctor's face making it difficult for Pong-Yul to read his disposition. In any case the Colonel was not terribly interested.

"You have examined G-101?"

"Yes."

"And…?"

The Doctor shrugged. "Dehydration, Lowered blood pressure, loss of blood, some disorientation. He's in shock of course from the beating and the bullet wounds."

"How long will he last."

Dr. Ho-Sik lit a cigarette, he did not offer one to Colonel Pong-Yul. Ho-Sik took a first deep drag and when he quit pulling Pong-Yul watched the smoke feather up into the darkness in lazy Ss. The Doctor spoke resignedly. "Cell deterioration has accelerated. Over the past few weeks he has expended a tremendous amount of energy and he's now on his last legs. In his present state I do not think he can heal himself..."

Pong-Yul tapped his foot impatiently. "How long, Doctor?"

"Perhaps a week, ten days at the most."

Pong-Yul swaggered to the door. He snapped his words back over his shoulder, his tone icy. "Now that I have him back I will not take the chance of losing him again. I wish to see him in ten minutes for interrogation and thereafter I trust you will sedate him—with a needle this time." Ho-Sik nodded his head and Pong-Yul stepped into the hallway.

CHAPTER 39

Peter loped north along the rocky ridge to intersect Trail #3. "Mr. "A". Have you figured out how I'm going to get down into the valley without the Koreans knowing?"

Jason came back his voice indicating his satisfaction. "Christa and I are working on it. Select a spot on the edge of the caldera where you have a view of the floor. Remember they have quite sophisticated detection methods so be discreet. I'm recalling Condor #1 and replacing it with Condor #2, fuel is running low"

* * * * *

Peter intersected Trail #3 and moved quickly but cautiously up the side of Voodoo Mountain. Cray gave the trail ahead of him and around him a clean slate with no sightings of unfriendlies so he took his time planting sensors on his back door. When finished he slid into a copse of pines nestled among the rocks on the edge of the caldera and moved up to where he had a good, clean view of the valley floor.

Shadows were already making their way across the valley floor from the east.

He set up his equipment and was comforted when the Condor showed no activity on the valley floor below him.

* * * * *

"Commander."

"Yes?"

"Motion sensors are picking up a large target, a bear, a man or a deer, on the edge of the caldera in Sector C."

"Put it on infrared scan."

"I have it Sir but because of the thickness of the underbrush I'm only getting partial information."

"How long will it take you to complete your scan and have something usable?"

The technician consulted his screens, "Perhaps as much as two hours."

Colonel Pong-Yul checked his watch. The caldera will be in deep shadow in two hours, he thought. But there will still be sun up there for another three. Time enough to do the job. "Keep me posted." he said curtly and strode through the door.

* * * * *

Peter had crawled into the copse and pulled some of the bushes out by the roots to reduce his cover in the front to just a little over twelve inches. Time became boring and he spent it alternating between watching the polymer screen and picking up his binoculars for a live look. The Condor hovered almost a thousand feet above the floor of the caldera and some nine hundred feet below him but he could not make it out against the shadows of the valley floor. He finished clearing a spot large enough for himself and stood briefly to shed his pack. Cray spoke instantly. "Peter, sit down."

Peter dropped to the ground. "Cray, what's the problem?"

"You are being scanned from below by infrared. They now know your weight and that you are bipedal."

"Damn!—I forgot."

"To confuse any further information you should lay with your torso toward the beam. It will be much harder for them to read then."

"Cray, is there any way of knowing how much they got of me?"

"No, Peter. However I would have taped the encounter and analyzed the tape afterwards. They would be in that process."

"Cray, any idea how long that might take them?"

"Judging from prior encounters with their expertise perhaps three minutes."

Meaning they've already got my number mused Peter.

* * * * * *

"Commander!"

"Yes."

"Colonel, Sir. I have data now on an unidentified, unauthorized human observer in Sector C."

"This is the target you informed me of earlier?"

"Affirmative."

"How far is he from here?"

The operator scanned the green digital figures in front of him. "Seven hundred eighteen meters, Sir."

"I will be right down."

"Yes, Colonel Sir."

Pong-Yul banged his way through the door into the Command Center and asked. "Do you have him on infrared scan?"

"Only partial now, Sir. However he stood up for just a moment and turned around and I was recording. I analyzed the recording."

"Can it be anything other than human?"

"Negative, Sir."

"Description?"

"Definitely human sir. Large male, too big for female, probably not oriental for same reason. He's in a pine thicket on the edge just where he can see us. When he stands he is approximately two meters high and 100 kilos. He's well concealed by the bushes but the reflection reading I'm getting on his kilocalories indicates he is Caucasian but must be wearing smut on his face, and, I think a pack on his back. His back pretty well blacks out when he turns around."

The Commander suppressed a smile. This was impressive information to come from a computer screen. "Any weapons?"

"Yes Sir. One rifle, eighty centimeters long. In extreme close-up mode I can pick up an advanced red-dot laser scope. See it there. He's definitely a sniper." The technician pointed to a faint point of light on the screen which was splitting the bushes two thousand three hundred feet above them on the brow of the caldera. "He's scanning us again down here. I can home in on that dot and zap him."

Pong-Yul smiled. "No, I don't want to waste a falcon missile. In view of his technology he's not local." He turned to the Watch Commander. "Lieutenant, I want to talk to him. Dispatch a squad up there and bring him to me—alive."

CHAPTER 40

"Peter!"

"Yes, Mr. "A"."

"A four man squad just exited the building and they're making straight for the north gate. You should be able to see them on your screen."

"Yes, I see them."

"I have a feeling they are a reception party for you."

Peter shifted and eased his binoculars through the weeds. Four ant like figures were just moving through the gate. As he watched they went straight to the base of the path and began to climb at forced march.

Jason's voice came to him calm and reassuring. "Will be at least half an hour before they get anywhere close to you. In fifteen minutes I'll fly a second Condor so we can keep track of them."

"Thanks."

"In the meantime, Christa has been unable to locate any plans for the compound. There was no requirement for them to file more than

219

the general aim of their construction so they didn't. All they said was they were building a laboratory and living quarters for a small staff."

"They could have a munitions depot under there."

"Precisely. However, Christa has uncovered some of the suppliers, notably bottled water and plumbing supplies, particularly toilets. Best estimate from the number of toilets delivered gives us seventy five people. Of course there's a healthy margin for error."

"The bear doth truly shit in the woods."

Jason laughed. "Aptly put. Also the compound is enclosed in green, six foot chain link fence carrying a man-killing ten thousand volts, hence no guards."

Jason paused and pulled a piece of paper from the stack in front of him. "Peter, some very disturbing thoughts about the WoodChopper. Relax and just listen for a moment. Christa and Cray have been sifting literally thousands of bits of data and one of those communiques was from the commander and he gave the doctor's report on The WoodChopper. Number one, they consider him a deserter and death is the antidote when a deserter is caught so he may be dead already. Number two, according to this report he was the object of some kind of genetic restructuring and in his case it went bad. It was supposed to allow them to rearrange the surface molecules of his flesh to any configuration they chose.

I think maybe what they are after is someone who can look at you in the flesh and then recreate you like a clone. He was one of the early ones and the process went awry. It left him with molecules that will not stay in place so his face and head have relaxed into the path of least resistance."

Peter interrupted. "My God, what a hare-brained idea. For what purpose?"

"Peter we live in a society in which law, religion and thousand story buildings are all artifacts spawned in man's mind, so here's just another wee step for mankind. This has to be for espionage reasons, what else? We don't know the scope yet but I'm hoping we will before they do any national damage. They've obviously been interrogating The WoodChopper since getting him back and we've been monitoring their transmissions. He killed the boys because they surprised him while he was asleep and he couldn't avoid them. It took him a day and a half of back breaking work to move them to the crevasse."

"Bastard! But how about the other two, he didn't have to kill them."

"He's having some kind of seizures and doesn't think rationally in the middle of it, otherwise he probably would have waited till later to enter the store. Kathleen was so horrified by his appearance that she was dead before she realized what was happening. He peed on them all to show his contempt for his superiors and mankind in general."

Jason put the paper back on the stack and studied the large wall screen. It was split in the middle to show pictures from both Condors. On the right Condor #2 clearly showed the four blips moving up the path. He shifted to color reception and they were clearly visible coming over the top and double timing it along the perimeter path. As he watched they dropped over the edge into the cover of the woods. They were less than a mile from Peter.

"Do you see them Peter?"

"Yes."

"You'd best busy yourself getting ready for them. Incidentally, now more than ever we need G-101 if we can get him. If nothing else we need a piece of him, something we can put under a microscope."

"With pleasure, any particular part."

Jason ignored the jest and said, "Facial tissue preferred but a finger would do."

Christa made a sound of disgust in the background and Peter rejoined. "Do I hear a note of disapproval from the lady of the house?"

Christa's voice came through his ear piece from faraway. *"You two sound like ghouls 'divvying up' body parts.*

Peter was about to retort when Cray cut in. "Peter, you have four probables Northwest at maximum range. ETA your position in twelve minutes."

"Thank you Cray. Gotta go Mr. "A"."

"Good luck Peter." Jason and Christa said in unison.

Peter slid back through the copse and sprinted down hill through the trees and heavy brush, pine needles cushioning the sounds of his rush. Two hundred yards down the hill he melted into the outer arms of a large rhododendron and took up his vigil directly in the path of the squad as they would journey up the slope.

All four were plainly visible on his wrist screen and in less than a minute he had visual contact on the point man as the squad began its surreptitious move up the hill toward his hiding place. The four men were spread out covering a swath of perhaps thirty yards.

He laid the Anschutz on a comfortable spot on the top bows of the bush and brought the red dot to the forehead of the soldier farthest away from him. He stopped breathing and squeezed off a shot. Before the first had clattered to the ground he had sighted on the second and dropped him before he could react to the danger.

Two down—two to go. The squad leader who was only twenty five yards below Peter dropped to the ground and began a quiet chatter into his radio. Peter moved silently across his front, traversed down the hill twenty yards and jerked the radio off the nearest soldier. He pressed the radio against his wrist scanner and whispered, "Cray, translate visually."

"Yes, Peter."

The screen filled with words.

"Number 3 answer, Number 4 answer—Number 2?"

"Yes leader?"

"I get no answer from 3 and 4. Do you see them?"

"No."

"You are closest to them, move over that way and observe."

"Yes, Leader."

Peter watched Number 2 reluctantly detach himself from the relative safety of a clump of trees and move silently to his left, angling up the slope. Number 2 had the misfortune of having to cross a small glade ten feet from Peter and Peter shot him just behind the ear. The dead soldier hit the ground with a small thud. The sound just reached Number 1.

"Number 2! Number 2! answer me!" came the agonized whisper.

Peter grinned, put the microphone to his mouth and breathed deeply and rapidly into it. The squad leader listened for a moment than whispered frantically into the radio. Peter put his radio back alongside the chronograph.

"Base, come in base."

"Base here, what do you have to report?"

Number One was desperate. "This is Six Leader. I've lost contact with my three men, they've just disappeared."

The Watch Commander grunted, the men did not appear to be drunk when they left. "Don't be absurd, dog Corporal. You can't just lose three men like that. Have you reached your objective?"

"No, he's one hundred fifty meters up the hill."

"Six Leader if I report this to the Colonel you'll be shot. Collect your men and get on with it. Your orders are specific, you are to bring him back alive. Get moving."

Peter watched a forlorn Six Leader pick his way across the hillside to where he had last seen Number 2. As he passed into the glade and stopped to look down at Number 2 Peter ended his misery.

Peter started back up the hill with the Korean radio crackling with the anxiety at Voodoo Base.

"Six Leader, Six Leader, SIX LEADER, come in. What was that noise? Come in Six Leader, come in Leader. Commander, I've lost contact with Group Six."

"Group Six? They were bringing in the unauthorized observer weren't they?"

"Yes sir."

Peter slid back into the copse and took up his vigil.

The conversation from Voodoo Base continued, the radio operator had forgotten to close the key. "Have they made contact yet?"

"Negative, and I can't reach any of them."

"What is your scan showing, can you still see him?"

"I can still see the laser spot, but I haven't seen him on infrared in over an hour."

Pong-Yul stood lost in thought. I don't like this, I don't like this one bit. Something is radically wrong up there on the edge of the caldera and I've got to stop it. He turned to the Watch Commander.

"Lieutenant, give the Sergeant Major the coordinates. I want to drop a Falcon into that copse as soon as it's ready to fire."

Peter stared at the screen and reread the words and it suddenly hit him. They were sending him a missile and he had just moments before it arrived. He grabbed gear indiscriminately, loaded everything into his arms and exploded out the back of the copse as Cray said, "Peter, a missile platform is rising on the valley floor. Ignition expected in ten seconds."

Peter sprinted twenty yards to the side and dropped down behind a massive rock. He could see into the valley but the floor was covered in shadow. A small, pure white explosion lit the valley floor momentarily like a short term flare and a red line arched to the copse and the copse ceased to be. The explosion was deafening and reverberated from crag to crag and beat about his ears.

"Peter?"

"Yes, Mr. "A""

"All right?"

"Yes."

"How about your equipment?"

"I got all but a bottle of water."

"Good. While you've been out there picking up your Vitamin D I've been adding a new wrinkle to our equipment."

Peter was still concentrating on the valley floor even though it was now in total shadow.

"Look at your Chronograph."

Peter looked at his left wrist. "Well I'll be damned, hello there Mr. "A"."

Peter was looking at a two inch high color image of a smiling Jason on his screen. Jason was laughing. Peter asked, "Are you in real time with this thing?"

"Of course, just like phonovision." Jason's face turned grave. "We've just intercepted a message that is different than all the rest. The Koreans have been changing code by the message to throw us off and Cray has been doing a yeoman's job keeping up with them. Whatever they're doing up there is coming to fruition in the next twenty four hours. But that's all we have."

"Is that all they said?"

"No, they've changed the code to something we've never seen before. Looks like cuneiform, probably from early Mesopotamia. Cray is screaming on this one."

"Will we get it in time?"

Jason shrugged his shoulders. "Who knows, Cray is already back to the fifth century and still no luck."

"The only way for me to short circuit this thing is to get into the compound. How do I get down there?"

"Aboard a Condor, or rather underneath a Condor."

"Explain please."

"The Condor will carry over four hundred pounds. I've already modified one and can have it to you in fifteen minutes. Underneath it is a sling like a hang glider rig. You let it hover, you climb in and tell it where to go. When you let it go I take over and get it out of there. when you get ready to come out I send it in to you."

Peter chuckled. "I'm glad you thought of the return trip. I'll go in just after sundown over the far end of the compound. The fence ends at the face of the caldera on that side."

"Fine, curl up somewhere and take a nap." Jason said. "They haven't discovered the Condor yet so we can watch over you."

CHAPTER 41

"Commander."

"Yes?"

"I'm intercepting some odd signals from the satellite."

"Explain."

"I've checked the incoming coordinates. They are not addressed to us but they are being beamed directly at us."

"Can you decipher them?"

"Partially. They are in binary and they seem to be very detailed flight instructions."

Commander Pong-Yul looked up from his desk. "Is there anything above us? Have you scanned with Radar and infrared?"

"Yes, Sir, and nothing is up there."

"Well, keep monitoring and if you catch something meaningful, let me know. Probably slop over from the satellite."

"Yes, Sir."

CHAPTER 42

"Peter, wake up."

"Yes, Cray."

There was a faint click and Jason said. "Peter, Jason here."

"Hello, Mr. "A"."

"I trust you slept well. We have no time to waste. The Koreans are moving out by midnight. They're scared to death that something is afoot they can't control and they don't want the whole mission in jeopardy, so we have to get in there and stop them."

"I'm ready."

"Good, keep only your hand weapons and your communications gear and your flash. Shuck your camo suit and conceal the rest of your gear at this site. Then drop down the mountain a few yards out of their infrared range and move straight west for about two hundred yards until you come to a small glade. The Condor is there now waiting for you. Cray will put it on infrared so you can find it. When you get there come back to me." There was a click and Amador was gone.

Peter kept the flak vest on but stepped out of his camos with great relief. They had repeatedly saved his life but were hot in the wearing and presented many snags on which to catch equipment. He strapped the Anschutz to his leg, checked his ammo belt then inserted the screen, his pack and the remaining loose gear in the camo top and threw the whole wad into the middle of a large rhododendron bush.

* * * * * *

"Colonel Sir."

"An FYEO[12] has come in for you."

"Send it to me."

Colonel Pong-Yul opened the door and saw the stiff backed orderly coming down the hall. He glanced at his watch and waited. The time was 21:05 hours. With a salute, then a deep bow the orderly handed the Colonel the message.

As he decoded he steeled himself to stop the trembling of his hands. The time was so near and they had worked so hard to have anything mar the program now. The message was from The Supreme Commander of the Military Affairs Commission of the Central Committee, General Kye Yong-Sop. Decoded, it read:

"All systems are in place for your arrival in Virginia. Be airborne by 0500 Zulu hours. Destination coordinates will be supplied en route. Destroy all evidence, including G-101, by fire and explosion within thirty minutes. Our fate and the glory of our people is in your hand. End."

Pong-Yul pushed a button on his desk radio. "Watch Commander. Tell my Security Officer and Doctor Ho-Sik I want them in my quarters in five minutes."

* * * * * *

Peter made his way through the velvety darkness just below the top lip of the caldera. He skirted rocks and clumps of trees and rhododendron. His heart pumped in anticipation. His wrist scanner showed a red blip which steadily drew closer as he jogged. "Cray, is that blip the Condor?"

"Yes, Peter. It is now forty meters ahead of you."

Peter paced the meters and stopped as he came to the edge of a small clearing. He heard a sibilant whirr and a faint beep about ten feet overhead. He pointed a hooded flash in the air and was astonished to see a Condor hanging there, contra-rotors quietly whirling, its three foot platform rock solid, a web harness dangling from its center. From one side of the central shaft a long lever ending in a bicycle hand grip arched over the edge of the platform and down so that it could be grasped at the shoulder height of the operator. The whole apparatus was suspended so solidly in midair that it might have been attached to something above it but he knew it was not so. The Condor was a Jason Amador work of genius and would transport him into, and hopefully out of, the hell of Voodoo Valley.

"I'm here, Mr." A"."

"Peter, point your scanner at the Condor. This will give me the exact height off the ground."

Peter did and the Condor began settling immediately. It stopped three feet over his head. The web harness and flight stick dangled in front of him. Jason's voice came again through the darkness. "Do not touch the flight stick until I tell you. The harness is standard rappel gear and you know how to put that on. When you are in you should have about two feet between your head and the platform and the flight stick should be right off your left shoulder. That leaves your gun hand free. Let me know when you are in"

Peter shrugged and tugged, pulling webs here and there to make a snug fit. "I'm in."

"Good. The flight stick does three things. First, when you touch it you have command of the Condor. It will remain under your command as long as you have the grip in your hand and it will stay in the current mode until you move the stick. Second the bicycle hand grip rotates. It is the throttle and it operates just like the chopper throttle. You are at idle speed now. When you want to ascend you twist the handle and increase the speed of the rotors. When you want to descend you decrease the speed. Twist the handle a quarter rotation"

Peter gingerly squeezed the handle and rotated it ever so slightly. There was a discernible decrease in rotation above him and the web loosened. Quickly he reversed the handle and the web tightened into his crotch. In a moment he was balancing on tip toes. He reduced the speed slightly and the Condor settled back and released its pull on the harness.

"Its very sensitive."

"Yes, you'll just have to be careful till you get the hang of it."

"How much testing has gone into this rig."

"How long have you been in harness?"

Peter stopped for a moment and Jason's retort suddenly sunk in. "You mean no one has ever flight tested this thing. Gimme a break, Mr. "A""

"Peter I have sublime faith in your ability and my know-how. Trust me, it will work."

"Humph! How about direction?"

Jason was smiling and was glad Peter couldn't see him. "The flight stick swings three hundred sixty degrees. Incline the blades in the direction you want to go. If you want to go forward you pull the stick toward you. For reverse, push the stick away from you. If you let go of the stick the gyros will pull it back to level flight. Cray says top speed on a level will be eighty knots dragging your reduced weight."

"How about turning it over?"

"Can't be done, the gyros are controlled by a chip and they won't let it happen as long as you have power."

"And if I lose power?"

"The gyros are revolving around eighty thousand rpm.[13] and will give you at least a minute of stability."

"Yeah, but I'll be coming down like a stone."

"Immediately throw your stick into neutral. If you're over fifty feet from the ground you'll auto rotate down."

"And if I'm under fifty?"

"You'll still auto rotate but the mechanism won't have ample time to fully compensate. Brace yourself for a hard landing about like you'd experience jumping out of a second story window. Remember in all power out cases you must be ready to shunt the platform off to one side of you as you land."

"How much does it weigh?"

"Eighty two pounds."

"Piece of cake," Peter grinned in the dark.

"Peter, elevate yourself a couple of feet and move back and forth in that clearing. Get the feel of it when you are just a bit off the ground. With your experience it should be relatively simple. Remember that as soon as you put your hand on the grip we no longer have control."

Peter played with the contraption and found it a pilot's delight. As sensitive as a safecracker's fingertips it responded to his slightest nudge and in five minutes he felt like he had been flying it forever. He idled it down and suspended himself an arm's length off the floor of the clearing. He deliberately swung his weight against the straps like kids do to get a swing started. The platform resisted any movement off the status quo.

"I'm ready, Mr. "A". I want to go in from the top of the main building."

"Peter there's a flat hatch in the southeast quadrant of the roof. It appears locked."

"How do I get through?"

"Reach up above your left shoulder on the top side of the platform. Careful, don't reach into the blades…"

"I feel something strapped down."

"…Yes. Up there are five, ten round magazines of hand-loaded HE[10] ammo for the Anschutz. Also a tiny oxyace[11] cutting torch. It'll give you about fifteen minutes at twenty-four hundred degrees which should be enough time to cut the hinges if you need. It's quieter than explosives, although I think they'll be aware of you right from the start. There's also an energy pak. Unwrap it and chew it down slowly as you are flying into the caldera. Also up there is a pair of Russian low light night glasses with a line attached to the platform. They're a little bulky but they magnify light some fifty thousand times. Makes night almost into day if you don't mind green day. They should come in handy going in."

Peter pushed the magazines into his belt and put the glasses on. The world immediately turned green but the trees and rocks around him stood out in near daylight relief. "Mr. "A", you think of everything."

"We try. Thank Christa for remembering your energy level."

The phones went dead for a few moments while Peter flexed inside the web making last moment adjustments—then Jason came back again.

"Peter, it's beginning to look like we have an incident of National importance here. What kind of monstrous skullduggery has been hatched down there is beyond me but I do know it is of the greatest necessity that you abort it. It's no use me telling you what to do once you get there. Just lay waste to their plans Peter, lay waste. Everything down there is expendable. Try to bring out some kind of proof of their intentions. Good luck. I won't bother you from here on unless we get information that you need or you call us. God Speed."

Peter breathed a silent Geronimo and lifted above the tree tops. Hanging below the platform was like being suspended in the middle of the world. Time stood still as the panorama of dark green crags surrounded him. The night glasses did not reach to the valley floor so dropping into the caldera was like dropping into the deepest recesses of a greenish black ocean. He munched on an energy pill as he descended. He dropped slowly at first then more rapidly as he gained expertise with the controls and controlled his pumping heart.

At two hundred feet the valley floor came into view and Jason came on line. "Peter, hover here momentarily and turn your scanner to straight color. They have one camera rotating three hundred sixty degrees mounted fifty feet above the northeast corner of the main building. It scopes the compound once a minute like a radar screen. You are above it right now. Watch it's rotation and slip in on the blind side and a three second zit of oxygen from the torch will freeze the

lens and break it."

"Gotcha, Mr. "A"."

CHAPTER 43

"Commander!"

"Yes Lieutenant."

"We have just lost our compound camera and I have a blip on infrared."

"Can you identify it?"

"Yes, Colonel Sir. It is a man and his readings are identical to the one up in the copse. He is on the main roof."

"How did he get up on the roof? None of our motion sensors have tripped and nothing came through the mine field."

"There is other infrared information with him, Colonel Sir, but I cannot make it out. It is a small heat source—and now it is gone."

That last explanation was lost on Pong-Yul who was deep in thought. So we didn't get him. Where did he come from? Well providence is kind, now's my chance again. "Where is his most likely point of entry, Watch Commander?"

"The steel door on the roof is locked, Colonel Sir. I would guess he will come to ground and try one of the ground level doors."

"Put two men inside each door immediately to wait for him. Let him in. I want him alive if possible. Do you understand?"

"Yes, Colonel Sir."

* * * * *

"Peter."

"Yes Cray."

"You are being scanned by infrared from nine o'clock, forty meters."

Peter had just released the Condor and was crouching on the roof. He brought the Anschutz up and probed with the laser dot. In a moment he crossed the infrared beam and homed in. A single silenced shot canceled the infrared scanner in a shower of sparks.

* * * *

"Colonel sir! We have just lost our infrared scanner. I have no eyes to watch the intruder."

"Don't worry he'll be coming down the outside ladder about now. Put two men up through that hatchway onto the roof. We'll flank him."

* * * * *

Peter sprinted across the roof to the hatchway. It was raised slightly above the roof and was made of solid steel with three hinges along one side. Large counter weights were slung in the air on brackets. At least it would be easy to open when unlocked. He ticked it with his finger and heard nothing in return but a dull thump. It would be a shame to get this far and be foiled by a simple steel door, he thought. He laid down beside it, ignited the torch and let the flame lick the first hinge.

He heard a sound within, and shut the torch down. Handles were being turned below the door. As he watched the door began to rise on oiled hinges. He rolled silently to a crouch at the end of the door. The door opened fully and a head appeared but looked the wrong way first. Peter placed both hands around the emerging soldier's neck and levered him up and out of the hatchway. It was a simple twist of his huge, muscular hands to snap the child sized neck and lay the twitching body quietly on the roof. There was a call from below. He grunted an answer, stepped away from the hatch into the darkness and sighted the Anschutz down into the opening. As the second soldier stood uncertainly at the bottom of the steps Peter dispatched him with a single shot.

He looked down through the steps. They came up from a small room. It was empty. He dropped through the hatch, quietly closed it and flew down the steel steps into a storage room. The only door led into the hallway. He peeped around the door. A bell rang insistently and a red light flashed at the end of the corridor. Beside the red light a security camera panned the hallway like a one-eyed owl. He put the laser sight on the lens and pulled the trigger. The camera sputtered and disintegrated in a pall of black smoke.

He slid quickly into the hallway. Several doors led off into separate rooms. He clicked the Anschutz to rapid fire and advanced stealthily to the first door. With his blood pumping fire through his veins he popped it open. Nobody there.

The next and the next and the next were empty. He paused outside the next to last door and heard a rumble of voices raised in panic. He slammed his foot against the door and burst into the room.

Twenty or so people were seated at the long boardroom table and were equally stunned by his entrance. The tableau held for one long breath in which he mentally counted fourteen civilian Caucasians, one high ranking North Korean Officer, one NKPA non-com and two orientals wearing long, white hospital smocks.

Peter's mouth flew open. He was looking directly at the President of the United States. He threw up his arm and scanned the room. As he did so, the crowd broke as one and flowed in a panicked lump for the door at the end of the room. The President was followed closely by the First Lady and the Vice President.

Not every one reacted in panic. The non-com drew his weapon and opened fire. Peter grunted in pain and shock as two direct hits took him off his feet and dumped him into a heap in the corner behind the door he had just entered. On all fours the Senior Officer scrabbled out the door at the other end of the room but the non-com pressed the advantage and advanced on Peter, firing wildly. Instinctively Peter leveled the Anschutz and stopped him in his tracks with a burst.

CHAPTER 44

Jason and Christa were stunned as the scanner fed the bits and bytes through the satellite and Cray reassembled them on the big wall screen. The room was in pandemonium. Ten seconds after the pictures hit the screen Cray was keying names under subjects.

"I do not believe it. The entire United States Cabinet," Jason gasped. "They have duplicated every important figure in American Government. All they have to do is replace them and they'll own us!"

On a second screen Cray began printing out messages as they were being decoded between Voodoo Valley and Japan. Jason stared in disbelief as he read the words. He screamed into the radio.

"PETER! My God I should have tumbled to it. Cray has identified the full United States Cabinet in that room. We're picking up panic transmissions back and forth now and here are the pieces. The genetic experiments have been to create look-a-likes for our heads of state, including the president so they can replace our people with these people. It is mind boggling. This would bring the United States totally under the control of North Korea. STANDBY!"

Peter was dimly aware of getting to his feet. He took a deep breath and checked his damage. One of the 9mm slugs had caught him in the meaty area high up under his left arm and spun him around. The other had left a dent in the Kevlar vest just over his heart and had knocked him down. His whole left side was beginning to throb with every heart beat and hurt like hell. Blood covered him from his armpit to his belt and dripped from his fingertips. He ignored it and went to the door through which the crowd had disappeared. It was locked but yielded to a well placed burst from the Anschutz. He peeped out the door. The hallway on the other side was empty except for the ubiquitous camera. He lurched into the hall, dispatched the camera with a burst and staggered along the hall, trying to get his breath back, opening each door in turn.

Jason talked to him as he moved. "Peter, they've got orders to evacuate immediately by chopper. They're taking two birds but they don't say where except as soon as possible. They've given their personnel orders to shoot you on sight."

Peter came to a corner in the hallway, bent down low and inched his scanner around the corner. Two guards were fifty meters down the hall waiting for him. He popped a magazine of HE into the Anschutz and flipped the switch to automatic. He bent over, propped the muzzle across his toe, eased his foot around the corner and pulled the trigger. There was a deafening, staccato roar and screams from the other end of the hallway. A cloud of dust and small debris wafted back to him. He glanced around the corner and could see nothing but holes in the walls and bloody body parts strewn about.

He sat down heavily on the floor. Blood had soaked his whole left side down into his pants. Everything was sticky and the smell made him nauseous. The slug had gone clear through the meat but had punched a savage hole going out the rear. It was becoming impossible to ignore the pain. He flexed the arm and sucked in his breath. God, it hurt. The hallway swayed back and forth. I have to stay in charge, he thought. I have to get back on my feet.

He became aware that Jason was talking to him.

"Peter, I know you're hit. Can you make it back to the roof?"

"Yeah." said Peter weakly.

"Do so now, those choppers are airborne in less than five minutes. You must do your utmost to stop them."

Peter gathered his strength and sprinted along the hallway and around the corner and ran smack into four marauding soldiers. The quarters too close, and the surprise on both sides too absolute for gunfire. He waded into the four slashing, crunching and chopping and in fifteen seconds was on his way along the hall leaving a heap of broken backs, arms and necks. Somewhere in the melee he had taken a deep cut across his left forearm. Temporarily he ignored it but he left a second bloody trail as he brushed against the sides of the hallway. He entered the storeroom and pounded his way up the stairs past the dead guard. He opened the hatch and the stairway was flooded with the backwash from the Condor hovering on the rooftop. He cleared the stairwell, grabbed the straps hanging down and began to painfully work his way into the harness. He snapped the final catch and donned the night glasses.

Suddenly he heard a grinding noise and the roof began to vibrate. He looked over the edge and movement on the ground caught his eye. Trees were sliding away, a small building slid the opposite way and lights probed up into the night sky. A black hole was opening in the ground. From it he could hear the angry whine of turbines and the popping of whirling rotors as a big chopper clawed its way out of the black hole and slashed up into the blacker night.

Jason screamed in his ear. "Scan it Peter, scan it!"

Peter thrust his arm at the chopper like Thor loosing a bolt. The bird was close enough to see white faces pressed against the windows then it veered away. He could feel the turbines throb and the rotors complain as the pilot brutally forced his ship into maximum lift up and into the Dead Man's Curve[12] and out of sight into the inky darkness. Peter's Condor lurched momentarily in the furious down wash then the gyros compensated and it held steady above him.

Jason's voice was a comfort midst the deafening chaos. "There'll be a second one Peter, scan it!"

Even as he spoke to Peter the second chopper popped into view—spit out of the hole like a missile out of a concealed submarine. It came from the giant black hole like a hungry shark, slid backwards a few feet, rotated one hundred eighty degrees and chattered up into the stygian night.

"Peter! The whole group you saw in the room is on those two choppers. We've got to stop them! I know you're hurt—how badly? Can you fly?"

Peter grunted through the pain. "Affirmative. I've got a hole in my shoulder and I'm bleeding from a cut on my arm—but we're wasting time. I'm lifting off, I'll talk to you when I'm up."

He eased the throttle around, felt the harness dig into his legs, then rotated the throttle again and wafted up into the night. At thirty feet above the building he turned to maximum lift and shot up through the blackness. His whole left side was aflame, he was dizzy and nausea ate at him and in the velvet abyss of the caldera he began to lose touch with reality. The air that cloaked him was dark and dense and he could not tell if his eyes were open or closed. He was so deep in the caldera that even the stars were lost in the mouth of the volcano. Only the whir of the blades and the constant tugging of the Condor above him and Jason's occasional words of encouragement in his ear gave him a sense of direction. He spoke to Jason.

"Mr. "A" what are we going to do about the first chopper?"

"Christa is following it with another Condor and she's right on top of it. She's like a kid in an arcade playing video games. It has cleared the caldera and is heading North Northeast. She'll blow it out of the air when it's in the right place."

"Where's mine?"

Jason could feel the utter exhaustion in Peter's voice. Peter was near collapse. At whatever cost he had to stay conscious until the deed was done. Jason forced himself to stay calm and measured his words with great care. "Cray has it above you just about to climb out of the caldera. You can climb faster than he but when you come out you'll still be about two hundred meters behind him. The other chopper has dropped to tree top level and reduced speed to sixty knots. They obviously think they are safe now. If yours does that you should be able to catch him easily."

Peter threw his head back and looked up through the blackness. All he could see was the darker outline of the three foot platform above his head. It was so close to him that it obscured the stars overhead and left him without a benchmark.[13] He dangled in limbo like a fly in a spider's web, having no knowledge of either rising nor lowering, his mind and body caught in the numbing shock of the last half hour. He needed some connection to something. To force himself back to reality he asked an inane question of no one. "Why don't we just notify the authorities and let them pick these bastards up as they come down?"

Jason's voice came back to him in the blackness, a soothing father's kind of voice, a voice to trust, a verbal bandage to wrap around his battered senses.

"No time to establish our credibility, Peter and no time to mount a defense. We don't know where they're going to land, nor even when. If they land those choppers those clones will disappear into the scenery and we'll never know if the right ones are in Washington or not. No, my friend, we are the avenging angel and we have to stop them."

"How close am I to the top?"

"Fifty meters or so. Your chopper is heading after the other. Turn Northeast when you top out and you should be able to see their exhausts.

CHAPTER 45

Jason turned to assist Christa. She was seated at the console next to him, the long tapered fingers of her right hand closed gently around a joy-stick. Her left hand played a keyboard. Jason thought of a cyberpunk in a video arcade. Above her, on the wall, next to his screen was an identical six by six foot screen. Roughly in the center of her screen was the dim outline of a flying chopper. She was totally absorbed in her game of cat after the mouse.

Jason read the coordinates as they continuously changed at the top of her screen. He smiled. "They're roughly sixty miles south of Gatlinberg right in the middle of the National Nantahala Forest. That's very rough country and almost totally unpopulated. You could drop a ship in there and no one would find it for years. Kill 'em, Christa."

Gleefully, Christa elevated the speed of the Condor. The chopper's exhaust grew slowly larger. She pressed a key on the keyboard and a set of cross hairs appeared on the screen. She fingered the joy stick and the cross hairs sought out the red exhaust and steadied on it. She pressed a red button on the top of the stick. A thin stream of tracers stitched the screen and disappeared into the exhaust. The screen turned into one immense fireball and went blank. She turned away from the console. "Mission accomplished," she said quietly, "but I lost the Condor."

"It was expendable. Good work. Now help me get our friend home."

* * * * * *

Peter's vision was blurring and it was hard to hold his head against the sixty knot wind. His body streamed out behind his head, buffeted like a rag doll puppet bouncing at the end of the strings of an angry puppet master. Although numb his left hand was glued to the flight stick and he had it pulled far back in maximum forward attitude. The tiny Condor chewed its way through the night like a predatory wasp after a hapless moth.

The angry fireball that erupted half a mile in front of him thundered about his ears and the shock wave shook him back to life. He passed through the smoke of the wreckage and saw the diminishing fire beneath him before he had realized what had happened.

The second chopper in front of him went into immediate evasive action. The pilot twisted it up into the air, slid it off sideways and managed to miss the smoke and debris coming down but for a few moments the chopper faced him, flying backwards.

In panic he yelled to Jason. "I've lost it, I can't see the exhaust!"

Jason's voice came back quietly. "Patience, Peter. Check your scanner. Cray says it is still in the same place in front of you. They just bounced around a bit in order to miss the explosion. However, Cray says they are increasing speed. You've got to close in for the kill, now!"

Peter was flying so low an occasional leafy monarch reached from the floor of the forest a hundred feet below and brushed at his feet. He took the Condor to two hundred feet and the moon befriended him. There, one hundred meters in front of him, he could see the soft silver glint of the chopper blades as the pilot nearly swept the tree tops in his effort to remain undiscovered.

He turned the throttle to maximum and watched the trees sweep faster beneath him. The air was suddenly nice and warm and his head lolled off to one side as the distance to the chopper closed to fifty, then forty, then thirty yards. At fifty feet the chopper's exhaust became a tiny hole in his consciousness, a warm beckoning refuge and he flew toward it like a virile sperm carrying destiny to the egg.

"Peter! You are too close, don't fire at this distance. Peter! Answer me, do not fire at this distance, it'll blow you out of the air."

In Jason's screen the exhaust loomed large enough to swallow a football. He turned to Christa. "Christa, take the controls away from him!"

"I can't, he's got his hand locked on. As long as he's locked on I can't override him!"

The two watched in horror as the exhaust became an ugly rictus of searing flame and filled a quarter of the screen. "He's too close. He's less than thirty feet away from that damned ship. If he fires at this range it'll kill him!"

* * * * * *

259

Peter felt the warmth of the exhaust enticing him like the arms of a compliant seductress. Now was the time to fulfill the mission and lay his head against the soft breast and nuzzle.

"Peter!" Christa pleaded. "Peter! Please drop the controls. Give them to me. I've got to get you away from there before you blow that ship!."

Peter raised his head, someone was telling him it was time to blow it up. The exhaust looked close enough to stick his arm in. He pressed the switch for full automatic and raised the Anschutz in his right hand but couldn't hold the point. He let it drop wearily to his side then willed himself to lever it up again like the swinging arm of a crane. As he got it to eye level he brought his useless left arm over underneath the barrel to steady it at the instant of firing. He squeezed the trigger.

Christa jumped on the controls and the stick like a xylophone player playing "Flight of the Bumble Bee" then the screen erupted into a massive maw of fire and went blank.

Jason checked the data screen then collapsed into a chair. "Goddammit! He got 'em but we've lost him."

Christa mothered her controls. The stick still had pressure against it. She moved it experimentally. It caught and held.

"No, I've still got him!"

Jason looked at her incredulously. "How?"

She kept one hand on the joy stick and the other on the keyboard. She was sobbing with relief. "In order to steady his shot he had to put both hands on the weapon. Just as he fired I went for full descent. I thought I was too late but I've still got him. He must have dropped a hundred feet before the controls began to respond again. Everything aboard that Condor is knocked out except propulsion and guidance and he may be burned to a crisp but I've got him."

"Can you tell where he is?"

"Now I can," she said, indicating another screen. A small blip blinked at them.

"The GPS pulse is still working on his scanner."

"Yes."

Jason turned away to conceal his tears. "Bring him home, honey."

EPILOGUE

"How's the shoulder, Peter?"

"A-1, Mr. "A". Another few days and three or four workouts and as good as new."

Peter sat on the balcony of his apartment in Half Moon Bay twenty-five hundred miles from "L'Aerie" and smiled at the long, lissome bikini-clad lovely who was obscuring his view of the golden Pacific. "Your surgeon did a marvelous job and down the line I'll forget about it. How're loose ends there?"

Jason smiled. He was tipped back in a lawn chair, feet up on the railing on his deck above the pond where the trout lazily waited for his largess. "The sheriff invited me to go along so we choppered into Voodoo Mountain. The Koreans had left The WoodChopper there bound and gagged and the whole place was set to self destruct. However, they hadn't figured on his dying determination. He got loose and shut the system down. Everything physical was intact. Washington has him now and enough records to piece the concept of the research together."

Peter could hear Jason sigh. He continued, "Sadly though, they took everything with them that we could have used to embarrass them at the UN".

"How about "The WoodChopper? Won't he cooperate?"

"Dead men tell no tales—he died just after they found him."

"Shame. How about the debris from the two choppers?"

"After you and Christa got through with them not a chance. There's nothing out there big enough to identify from the air so they can't even find them. I've got the GPS coordinates but I don't relish their scrutiny if I give them to the bureaucracy, so I'm just going to let it ride. Oh, by the way—Christa says to enjoy your R&R."

Peter reached up and tweaked the dainty bow knot on the tiny thong that held the minute scrap of cloth around the tawny thigh standing in front of him. It dropped away. He sucked in his breath. "I'll try Mr. "A"," he said and closed the phone.

[1] Georgia Bureau of Investigation

[2] Knee high four or six wheelers with all wheel drive and one rough seat for the driver. Have a flatbed approximately six feet by seven. They are used to haul hunters/supplies/gear into difficult places and to bring personnel and dead game out.

[3] For Your Eyes Only

[4] Medical Examiner

[5] The Plus P ammo is the most powerful available in any caliber.

[6] Liquid Crystal Display

[7] An Amador invention, the Condor is a flying platform held aloft by two contra-rotating eight foot propellers. It is powered by a small, propane jet engine and is capable of over flies with a camera, weapons or a man suspended in the harness below. It can be controlled by home base or the man who is actually flying it.

[8] Meals Ready to Eat.

[9] Infrared Spectrophotometer: This instrument passes a narrow beam of infrared energy through a thin film of the substance being studied. As the wavelengths change, the amount of energy transmitted by the specimen is measured and recorded on a chart. The chart is a 'fingerprint' of the organic material being subjected to study -- a primary means of identification and comparison of plastics, rubber, paint, and other organic compounds. It has the advantage of being able to detect slight differences in composition and molecular arrangements of minute amounts of material, 4th Edition, Criminal Investigation, Swanson, Chamelin, Territo, Publisher McGraw Hill, Pages 233 and 234

12 [12]For Your Eyes Only

13 [13]revolutions per minute

[10] High Explosive

[11] A volatile mixture of oxygen and acetylene.

[12] The height-velocity diagram. The combination of altitude and airspeed from which structural damage to the helicopter will occur in case of a power failure; also called 'The Dead Man's Curve'. CF: "Learning To Fly Helicopters", R. Randall Padfield, Auth. Tab Books, Division of McGraw-Hill Publ. 1992 PP 165-9.

[13] A marked point of assumed or known elevation from which other elevations may be established. A point of reference.

About the Author

In his early years, Rus Morgan spent five years in the Marine Corps. Since then, he has been an account executive, a laborer, a poultry and cattle farmer and owned a construction firm. He is a Mensan who has published articles in the *Memphis Magazine*, *The Flyer*, *The Memphis Business Journal, Hands-on Electronics* and has published two other novels, *Blackberries Got No Thorns* and *Luci*. He resides with his wife in Memphis, Tennessee and can be reached through email at: RUSMOR4@aol.com or through his website at http://www.rusmor.com where you may inspect the full range of his other works and order autographed copies.